Wild Plant Species as Potential Horticultural Crops: An Opportunity for Farmers and Consumers

Wild Plant Species as Potential Horticultural Crops: An Opportunity for Farmers and Consumers

Editors

Roberta Bulgari
Ada Baldi
Anna Lenzi
Antonios Chrysargyris

MDPI

Basel • Beijing • Wuhan • Barcelona • Belgrade • Novi Sad • Cluj • Manchester

Editors
Roberta Bulgari
University of Turin
Grugliasco, Italy

Ada Baldi
University of Florence
Firenze, Italy

Anna Lenzi
University of Florence
Firenze, Italy

Antonios Chrysargyris
Cyprus University of
Technology
Limassol, Cyprus

Editorial Office
MDPI
St. Alban-Anlage 66
4052 Basel, Switzerland

This is a reprint of articles from the Special Issue published online in the open access journal *Horticulturae* (ISSN 2311-7524) (available at: https://www.mdpi.com/journal/horticulturae/special_issues/Wild_Plant_Species).

For citation purposes, cite each article independently as indicated on the article page online and as indicated below:

Lastname, A.A.; Lastname, B.B. Article Title. *Journal Name* **Year**, *Volume Number*, Page Range.

ISBN 978-3-0365-9670-9 (Hbk)
ISBN 978-3-0365-9671-6 (PDF)
doi.org/10.3390/books978-3-0365-9671-6

Cover image courtesy of Ada Baldi

Contents

About the Editors

Roberta Bulgari

Dr. Roberta Bulgari (Department of Agricultural, Forest, and Food Sciences, DISAFA, Vegetable Crops and Medicinal and Aromatic Plants, VEGMAP, University of Turin, Italy) is a fixed-term type A researcher at the Department of Agricultural, Forest, and Food Sciences (DISAFA), of the University of Turin, and is currently a professor of Horticulture—vegetable production at the University of Turin. She has obtained a PhD in Agriculture, Environment and Bioenergy in 2018 at the University of Milan, Department of Agricultural and Environmental Sciences (DiSAA). She specializes in vegetable and ornamental plant cultivation, vegetable quality, postharvest quality, soilless cultivation, greenhouse crops, leafy vegetables, wild edible species, microgreens, biostimulants, abiotic stress, and urban horticulture.

Ada Baldi

Dr. Ada Baldi (Department of Agriculture, Food, Environment and Forestry, DAGRI, University of Florence, Italy) holds a PhD in "Sustainable Pastoral and Forage System for Territory and Product Quality" at the University of Florence. At present, she is a Research Fellow at the Department of Agriculture, Food, Environment and Forestry at the University of Florence where she carries out teaching and research activities in vegetables and ornamental crops. Her main fields of interest are the relationship between plant and environment variables determining productive processes, soil disinfection by means of environmentally friendly methods, the soilless cultivation of vegetables and officinal plants, the nursery production of turfgrass and vegetables crops, new ornamental crops, urban agriculture, and food safety.

Anna Lenzi

Dr. Anna Lenzi (Department of Agriculture, Food, Environment and Forestry, DAGRI, University of Florence, Italy) graduated with first-class honours in Agricultural Science at Pisa University in 1991. In 1995, she achieved a PhD degree at "Scuola Superiore di Studi Universitari e di Perfezionamento Sant'Anna" (Pisa). Since 1998, she has been an Assistant Professor at the Department of Agriculture, Food, Environmental and Forestry (DAGRI) of the University of Florence, where she carries out teaching and research activities in the fields of Horticulture (Vegetables and Ornamentals) and Protected Cultivations. Her main research activity concerns the following fields of study: vegetables quality, with particular attention to nitrate accumulation in leafy vegetables; the hydroponic cultivation of vegetables, ornamentals, and officinal plants; microgreen production; soil disinfection by means of environmentally friendly methods; the characterization of ornamental crops; new ornamental crops; wild food plants; the nursery production of turfgrass and vegetable crops; cultural media; biostimulants; urban agriculture; and the microbiological safety of vegetables.

Antonios Chrysargyris

Dr. Antonios Chrysargyris (biologist), completed his master's degree in "Land Ecosystems/Biological Resources" at the University of Crete (Greece), and his PhD studies at the Cyprus University of Technology (CUT), at the Department of Agricultural Sciences, Biotechnology and Food Science, in 2016. He explored the intricate relationships between medicinal plants and their response to environments such as salinity stress and/or mineral unbalance. Currently, he is a researcher at the CUT, where he also teaches a series of undergraduate courses. His main research work is focused on medicinal/aromatic/ornamental plant cultivation and their mineral needs, nutritional value, and the evaluation of the biological activities of plant extracts and essential oils (antioxidant, antimicrobial, insecticidal, etc.). He also works on the introduction of unexplored plant species in intensive cultivation schemes (hydroponics and soilless cultures), evaluating innovative materials such as alternative growing media. He has published 109 articles in peer-reviewed journals and is listed in the TOP 2% of cited researchers in the world within his main subfield discipline (plant biology) for the years 2021 and 2022. He also serves as associate and guest editor in many academic journals (Heliyon, Frontiers in Plant Sciences, Agronomy, Horticulturae, Chemical and Biological Technologies in Agriculture, etc.) His interests include aromatic plants and vegetable cultivation; the development of strategies for plant nutrition and response to abiotic stress; soilless culture, substrates, and hydroponics; the postharvest storage and processing of medicinal and aromatic plants; essential oil analysis and biocidal activity; and the evaluation of natural products.

horticulturae

Editorial

Wild Plant Species as Potential Horticultural Crops: An Opportunity for Farmers and Consumers

Antonios Chrysargyris [1], Ada Baldi [2,*], Anna Lenzi [2] and Roberta Bulgari [3]

1 Department of Agricultural Sciences, Biotechnology and Food Science, Cyprus University of Technology, Limassol 3603, Cyprus; a.chrysargyris@cut.ac.cy
2 Department of Agriculture, Food, Environment and Forestry (DAGRI), University of Florence Piazzale delle Cascine 18, 50144 Firenze, Italy; anna.lenzi@unifi.it
3 Department of Agricultural, Forest and Food Sciences, University of Turin, L.go Braccini, 2, 10095 Grugliasco, Italy; roberta.bulgari@unito.it
* Correspondence: ada.baldi@unifi.it

By the year 2050, it is predicted that there will be 10 billion people on the planet, and along with this population growth, the need for food production will dramatically rise. Over the past few decades, modern agriculture has relied on intensive farming techniques in an effort to achieve high yields and meet the current food demand. However, future agricultural productivity improvements may not be sufficient to ensure such high demands. There are about 30,000 plant species that are considered edible; however, very few of them have been cultivated as food crops on a commercially significant scale. Wild edible species have the potential to lead food systems to be healthier, more sustainable, and resilient to the current climate change situation. Increased tolerance to several abiotic and biotic stresses, as well as high nutritional value and excellent nutraceutical properties, are common traits of wild plants, making them promising candidates as new horticultural crops.

This Special Issue covers research aspects regarding novel approaches for the outdoor/indoor cultivation of wild or underutilized species that can give new opportunities for growers to produce new food categories, particularly appealing to modern consumers. Cultivating wild species is also a way to preserve the ethnobotanical heritage and promote genetic diversity. Furthermore, the cultivation of food plants usually gathered in the wild could reduce the health-related risks associated with pollution and biological contamination.

Citation: Chrysargyris, A.; Baldi, A.; Lenzi, A.; Bulgari, R. Wild Plant Species as Potential Horticultural Crops: An Opportunity for Farmers and Consumers. *Horticulturae* **2023**, *9*, 1193. https://doi.org/10.3390/horticulturae9111193

Received: 18 July 2023
Accepted: 26 July 2023
Published: 31 October 2023

The domestication of new plant species is an ongoing phenomenon in agriculture, which is of great importance nowadays. Sulaiman et al. [1] reported that only 250 plant species are fully domesticated, while many wild plants could meet the challenge of producing crops that are resistant to a series of biotic or/and abiotic stress factors, such as pests or climate change. These authors collected and presented data from five case studies from five Eurasian regions with different climate conditions and ecosystems, including 20 taxa of wild species with a high potential to become novel cultivated plants. The presented species are culturally significant since they are employed in traditional cuisine and have promising economic value. Many of them are leafy vegetables (e.g., *Malva sylvestris* L., *Allium ampeloprasum* L., *Chenopodium album* L., to name a few) that are a great source of minerals, vitamins, proteins, and fibers; on top of that, they can be consumed raw (fresh) or cooked (boiled), and different plant parts can be used (leaves, flowering bulbs, fruits, roots, etc.). The cultivation of these species may supply rural areas, farmers, and consumers, with more diverse and healthy food resources, as they are adapted to different climate conditions. Future studies should focus on the agronomical performance of these species, whose cultivation would also be a way of protecting native populations, as some of them are considered vulnerable/rare.

Wild edible leafy plants are a rich source of nutrients, antioxidants and have special organoleptic features that are highly appreciated by consumers. Baldi et al. [2] cultivated four wild edible leafy vegetables, namely dandelion, sorrel, wild chicory, and wild lettuce (*Taraxacum campylodes* G.E.Haglund, *Rumex acetosa* L., *Cichorium intybus* L., and *Lactuca serriola* L., respectively) in a floating hydroponic system, to evaluate their performance from an agronomical, chemical, nutritional, and sensory point of view. Plants were grown for 7 weeks and harvested at the baby leaf stage. The analysis that followed revealed promising prospects for using these species as baby salads. Features such as yield and mineral content were comparable to those of already cultivated leafy vegetables. Higher yields were obtained in sorrel (3.2 kg FW/m^2) and wild chicory (3.7 kg FW/m^2), while the mineral analysis of wild lettuce revealed that this was richer in phosphorus (P), calcium (Ca), and magnesium (Mg), compared to cultivated lettuce at the same growth stage. The four species did not show health risks in relation to heavy metal accumulation. The evaluation of their sensory attributes revealed a quite distinctive profile, which could promote their value as consumers' demand for new, tasty, and healthy food increases.

Areas such as the Mediterranean basin are full of unexplored plant species that are commonly used for traditional medicine or traditional food recipes. One such species is *Sanguisorba minor* Scop., a wild edible plant that is commonly harvested in the wild in dry and semi-dry grasslands throughout Europe. This species can be consumed boiled, in soups, or in salads and is a rich source of amino acids, minerals, vitamin E, and antioxidants. Ceccanti et al. [3] compared and revealed the differences between wild and domesticated *S. minor* in terms of their nutritional value (minerals, free sugars, organic acids, etc.). The results showed that after the plant's domestication, there was an increase in a series of features such as protein content (18.80 g/100 g for the wild plants, 23.10 g/100g for the domesticated plants) and total organic acids (5.82 g/100g and 20.43 g/100 g, respectively). The cultivated plants reported a higher content of polyunsaturated fatty acids (50.90%) than saturated (13.40%), while both wild and cultivated samples revealed a low $\omega 6/\omega 3$ ratio (0.31 and 0.36, respectively), demonstrating the species' significance in the protection of oxidative and inflammatory processes. The authors concluded that, because of its excellent nutritional characteristics, *S. minor* could be used as a functional food or ingredient in the human diet. Furthermore, fertilization and domestication management may be a method to maintain or improve its nutritional profile.

The applied cultivation practices affect a series of parameters for each plant species. New cultivation techniques, such as soilless cultivations, provide new insights into plants' performance and attributes. Toward this direction, Chrysargyris et al. [4] evaluated the performance of *Portulaca oleracea* L. in a soilless system, using residues from the extraction of essential oils from medicinal plants (*Origanum dubium* Boiss. and *Sideritis cypria* Post) as a component in the growing media mixture for partial peat substitution (0, 5, 10, 20, 40% *v/v*). High ratios (>10%) of these residues in the growing media reduced plant growth in terms of fresh weight (FW), leaf number, and plant height while also affecting the plant's mineral accumulation and physiological (stress) condition. This resulted in an increase in the antioxidant profiles of the plant that, in some cases, appeared more than doubled (up to 2.2 times). Additionally, when residues of *O. dubium* were used in the growing media (from 5% to 20%), the total phenolic compounds of purslane were increased up to 30.2%, in comparison to plants grown in 100% peat. This study concludes that tested residues can be used in a growth media at low ratios (up to 10%), given that, despite the growth inhibition, plants appeared to have improved nutritional profile and antioxidant status. In this study, the authors also promoted the sustainability of the cultivation/production of medicinal plants, as they demonstrated that all the materials produced by these crops (fresh and dry mass, essential oils, and residues) can be used. The waste, in fact, can be reused for other cultivation, even increasing the quality features of the produced plant, in this case, *P. oleraceae*.

The ability of plants to survive or to perform well under stress and abiotic conditions through increased electrical conductivity (in terms of salinity) or the deficit irrigation regimes have lately attracted research interests due to the scarcity or the low quality of water for agriculture. Alexopoulos et al. [5] cultivated *Hedypnois cretica* L. and *Urospermum picroides* L. under three salinity levels (EC fixed at 2.0–2.2, 6.0–6.2, and 10.0–10.2 dS m^{-1}). These wild edible species of the Asteraceae family are commonly used in the Mediterranean area as food or medicines and have been referenced to be rich in bioactive compounds. The results of this study indicated the strong impact of salinity levels on the growth (leaf number, rosette diameter, plant fresh weight, etc.) and quality (mineral content, chlorophylls, total phenolic compounds, sugars, acidity, etc.) of both species while *U. picroides* appeared to be more tolerant to salinity compared to *H. cretica*. It is worth noting that the nitrate content decreased considerably even at moderate salinity levels, with a positive influence on the final quality of the edible leafy product. The water issue is also the work of Guarise et al. [6], who evaluated hedge mustard (*Sisymbrium officinale* (L.) Scop.) as a potential leafy vegetable for the fresh-cut salad industry production chain. For this aim, two populations of *S. officinale* (from Milan and Bergamo, Italy) were studied. Firstly, plants were grown for seed production, and then the produced seedlings were cultivated under 100% and 50% of the crop water requirement. The obtained results demonstrated that the species could be cultivated with a 50% of water reduction without decreasing the plant's yield (ranging from 22.3 to 40 g per plant). Interestingly, no effect was noticed for a series of quality parameters, such as the anthocyanin content (47 mg kg^{-1} FW) and total sugars (ranged from 7–10 mg kg^{-1} FW); however, as in the case of increased salinity mentioned earlier, total phenols were 25% higher in the leaves of plants grown under 50% of the water requirements in plants originating from both tested populations, as a response to the drought stress, which induced the production of secondary metabolites. This species may be appropriate for increasing vegetable output in geographical areas with limited water availability, or as an ideal crop for winter production when irrigation is limited to avoid the spread of fungal infections.

Abiotic or biotic stress conditions are not the main factors that affect plants' growth and quality. The fertilization plan is the first and one of the most important practices in commercial farming, and it greatly adds to crop yield and quality. In the study of Paschoalinotto et al. [7], *Scolymus hyspanicus* L., an edible wild Mediterranean species, was cultivated under different fertigation regimes in terms of different levels of nitrogen (N), phosphorus (P) and potassium (K), ranging from 100 to 300 ppm. Seven experimental treatments were used, including six fertigation regimes and the control treatment, where no fertilizers were added. The growth, nutritional, mineral profile, and chemical composition of *S. hispanicus* were evaluated at the end of the cultivation period. The results revealed that the moderate levels of K and P (200 ppm) resulted in a higher fresh mass yield (113.58 and 116.27 g, respectively), while the lowest value was recorded when the input of N, P, and K was at the highest tested levels of 300 ppm each. The applications of moderate K and P also produced plants with increased nutritional features, such as fiber and carbohydrate content. Nutrient solution management appears to be a very efficient strategy to increase the fresh yield of *S. hispanicus* without sacrificing the nutritional value. On top of that, low inputs of minerals such as K and P may improve the chemical composition of the species.

Fertigation management, the irrigation regime, and other cultivation strategies are species-specific, and in many cases, different management is required at different growth stages of a plant or different ripening stages of fruits. The study of Arena et al. [8] investigated the patterns of the accumulation of carbohydrates, organic acids, and minerals during different developmental stages of *Hexachlamys edulis* (O. Berg) Kausel and D. Legrand fruits, which is an underutilized native species from South America, is representative of the Myrtaceae family; the pleasantly tasty sweet and sour fruits of the plant are referenced for their high antioxidant activity and high carotenoid content. Additionally, the plant's leaf extracts are well known for their pharmacological properties, as they are used against bronchitis and coughs. For the first time, variations in carbohydrates, organic acids, and minerals in fruit's ripening stages were investigated. The rise in total sugars, together with the absence of changes in total organic acids with ripening, suggests a sweeter and less sour fruit meaning that this is a desirable product to be consumed. The results of this study pointed out that the optimal time for *H. edulis* fruit consumption, in order to reach the maximum flavor and a good amount of nutrients, is the ripe or overripe stage, and suggested that the fruits are a promising source of compounds with high nutritional value. In the same context, Jia et al. [9] investigated changes in the physicochemical characteristics, antioxidant content, and nutritional composition of *S. obovatifoliola* subsp. *urophylla* (Hand.-Mazz.) H. N. Qin fruits at different maturity stages. This plant, endemic to China, produces large edible fruits with soft pulp and sweet taste, and the results of this study provide information for the proper maturity stage at harvest of fruits with better quality, longer shelf life, and better market acceptability.

Historically, gathering wild edible plants was the only way to survive amid famines and chronic poverty. These days, wild edible plants are more than a trend, and their introduction in cultivation schemes can provide not only important knowledge about the quality of the plants themselves but also allow the best cultivation mode to be found to make their production commercially significant. *Heracleum sphondylium* L., commonly known as hogweed, is an herbaceous plant of the Apiaceae family native to Europe and Asia. This wild edible plant grows in many different ecosystems and has been used in traditional medicine or as a food ingredient. Matarrese and Renna [10] reviewed the studies on this species, aiming to spread its knowledge and prospects as a new cash crop, which could be cultivated as both a horticultural and industrial crop. This review further indicates that ethnobotany may provide agricultural guidance, as hogweed and, more broadly, all wild edible plants have the potential to lead to healthier, more sustainable, and resilient food systems in the context of biodiversity enhancement.

This Special Issue collects a series of new approaches to the importance of the domestication of wild edible plants. The effect of cultivation practices, a series of abiotic stress factors such as salinity and deficit irrigation, were also evaluated during their introduction into cultivation schemes. Additionally, the effect of the ripening fruit stage on the nutritional features and the level of bioactive compounds of unexplored plants was also examined. According to the studies presented in this Special Issue, wild edible plants are promising candidates for both pharmaceutical and agri-food industries for the introduction of new agricultural crops and the achievement of functional products. Scientists from various disciplines (agronomists, chemists, pharmacists, biologists, etc.) need to combine forces and work toward the full domestication of wild edible plants in order to enrich food availability and diversity, while at the same time, protecting native populations.

Conflicts of Interest: The authors declare no conflict of interest.

References

1. Sulaiman, N.; Aziz, M.A.; Stryamets, N.; Mattalia, G.; Zocchi, D.M.; Ahmed, H.M.; Manduzai, A.K.; Shah, A.A.; Faiz, A.; Sõukand, R.; et al. The Importance of Becoming Tamed: Wild Food Plants as Possible Novel Crops in Selected Food-Insecure Regions. *Horticulturae* **2023**, *9*, 171. [CrossRef]
2. Baldi, A.; Truschi, S.; Bruschi, P.; Lenzi, A. Preliminary Assessment of Four Wild Leafy Species to Be Used as Baby Salads. *Horticulturae* **2023**, *9*, 650. [CrossRef]
3. Ceccanti, C.; Finimundy, T.C.; Barros, L. Nutritional Value of Wild and Domesticated *Sanguisorba minor* Scop. Plant. *Horticulturae* **2023**, *9*, 560. [CrossRef]
4. Chrysargyris, A.; Louka, S.; Petropoulos, S.A.; Tzortzakis, N. Soilless Cultivation of *Portulaca oleracea* Using Medicinal and Aromatic Plant Residues for Partial Peat Replacement. *Horticulturae* **2023**, *9*, 474. [CrossRef]
5. Alexopoulos, A.A.; Assimakopoulou, A.; Panagopoulos, P.; Bakea, M.; Vidalis, N.; Karapanos, I.C.; Rouphael, Y.; Petropoulos, S.A. *Hedypnois cretica* L. and *Urospermum picroides* L. Plant Growth, Nutrient Status and Quality Characteristics under Salinity Stress. *Horticulturae* **2023**, *9*, 65. [CrossRef]
6. Guarise, M.; Borgonovo, G.; Bassoli, A.; Ferrante, A. The Effect of Drought on *Sisymbrium officinale* (L.) Wild Species for Potential Cultivation as a Leafy Vegetable. *Horticulturae* **2023**, *9*, 111. [CrossRef]
7. Paschoalinotto, B.H.; Polyzos, N.; Compocholi, M.; Rouphael, Y.; Alexopoulos, A.; Dias, M.I.; Barros, L.; Petropoulos, S.A. Domestication of Wild Edible Species: The Response of *Scolymus hispanicus* Plants to Different Fertigation Regimes. *Horticulturae* **2023**, *9*, 103. [CrossRef]
8. Arena, M.E.; Povilonis, I.S.; Borroni, V.; Pérez, E.; Pellegrino, N.; Cacciatore, C.; Radice, S. Changes in Carbohydrates, Organic Acids, and Minerals at Different Development Stages of *Hexachlamys edulis* Fruit, a Wild South American Species with Horticultural Potential. *Horticulturae* **2023**, *9*, 314. [CrossRef]

9. Jia, T.; Feng, C.; Zou, S.; Gao, P. The Main Physicochemical Characteristics and Nutrient Composition during Fruit Ripening of *Stauntonia obovatifoliola* Subsp. Urophylla (Lardizabalaceae). *Horticulturae* **2023**, *9*, 29. [CrossRef]
10. Matarrese, E.; Renna, M. Prospects of Hogweed (*Heracleum sphondylium* L.) as a New Horticultural Crop for Food and Non-Food Uses: A Review. *Horticulturae* **2023**, *9*, 246. [CrossRef]

horticulturae

MDPI

Article

The Importance of Becoming Tamed: Wild Food Plants as Possible Novel Crops in Selected Food-Insecure Regions

Naji Sulaiman [1,*], Muhammad Abdul Aziz [2,3], Nataliya Stryamets [4], Giulia Mattalia [2], Dauro Mattia Zocchi [3], Hiwa M. Ahmed [5], Ajmal Khan Manduzai [6], Adnan Ali Shah [7], Abdullah Faiz [3,8], Renata Sõukand [2], Zbynek Polesny [1] and Andrea Pieroni [3,9,*]

1 Department of Crop Sciences and Agroforestry, Faculty of Tropical AgriSciences, Czech University of Life Sciences Prague, Kamýcká 129, 16500 Praha-Suchdol, Czech Republic
2 Department of Environmental Sciences, Informatics and Statistics, Ca' Foscari University of Venice, Via Torino 155, 30172 Venezia, Italy
3 University of Gastronomic Sciences, Piazza Vittorio Emanuele II 9, 12042 Pollenzo, Italy
4 Faculty of Forest Sciences, School for Forest Management, Swedish University of Agricultural Sciences, SE 73931 Skinnskatteberg, Sweden
5 Sulaimani Polytechnic University, Slemani 46001, Kurdistan Region, Iraq
6 Department of Environmental Sciences, COMSATS University, Abbottabad Campus, University Road, Abbottabad 22060, Khyber Pakhtunkhwa, Pakistan
7 Department of Botany, University of Peshawar, 2F4Q+42H, Rahat Abad, Peshawar 25120, Khyber Pakhtunkhwa, Pakistan
8 Faculty of Agriculture, University of Herat, Herat 3001, Afghanistan
9 Department of Medical Analysis, Faculty of Applied Science, Tishk International University, Erbil 44001, Kurdistan Region, Iraq
* Correspondence: sulaimann@ftz.czu.cz (N.S.); a.pieroni@unisg.it (A.P.)

Abstract: Domestication of new plants is one of the key (ongoing) phenomena in the history of agriculture. Wild plants are the ancestors of current and future crops and the largest reservoir of genetic diversity for crop breeding and improvement. Wild food species have been used for human nutrition since ancient times and are often the object of human strategies for coping with emergency situations, such as natural disasters and conflicts. We analyzed qualitative data collected through ethnobotanical field studies conducted in recent years in five selected Eurasian regions (Afghanistan, Kurdistan region of Iraq, Pakistan, Syria, and Ukraine) that have been recently affected by wars and/or socio-political turbulence. Data were collected through participant observation and semi-structured interviews with local people. We identified five taxa for each region, which are culturally very salient in the local food systems, that retain an important economic value in local markets, and that, therefore, could be good candidates for becoming novel crops. The cultivation of the reported species may significantly help local communities in their post-war livelihoods and especially in terms of food security and domestic nutritional care. Future studies should focus on the agronomic feasibility of the highlighted species within their regional ecosystems.

Keywords: ethnobotany; wild food plants; Afghanistan; Kurdistan; Pakistan; Syria; Ukraine

Citation: Sulaiman, N.; Aziz, M.A.; Stryamets, N.; Mattalia, G.; Zocchi, D.M.; Ahmed, H.M.; Manduzai, A.K.; Shah, A.A.; Faiz, A.; Sõukand, R.; et al. The Importance of Becoming Tamed: Wild Food Plants as Possible Novel Crops in Selected Food-Insecure Regions. *Horticulturae* **2023**, *9*, 171. https://doi.org/10.3390/horticulturae9020171

Academic Editors: Roberta Bulgari, Ada Baldi, Anna Lenzi and Antonios Chrysargyris

Received: 31 December 2022
Revised: 26 January 2023
Accepted: 26 January 2023
Published: 28 January 2023

1. Introduction

Domestication of new crops is one of the key events that formed the history of agriculture [1]. The number of plant species on our planet is estimated to be around half a million, among which, there are around 250 species considered fully domesticated, while other species are either semi-domesticated, undomesticated, or unidentified. Out of the 250 domesticated species, there are only a few species (e.g., potato, rice, wheat, corn, cassava) that form a source of 95% of the world's caloric intake [2]. Investment in these specific species puts huge pressure on the global biodiversity system; besides, undiversified food negatively impacts human health and leads to malnutrition. In addition to the role of wild plants in being ancestors of the current as well as future crops, crops' wild relatives remain the largest reservoir of genetic diversity for crop breeding and improvement, especially since 75% of the genetic diversity of agricultural crops was lost during the last century [3]. Wild plants serve in solving the challenges of producing crop varieties resistant to pests and diseases; they also contribute to finding new varieties adapted to the changing climate conditions such as drought, high temperature, and frost waves [4]. Several conventional and modern techniques are used in wild plant domestication such as selection, reproduction, hybridization, and genetic modification [5].

Wild food plants (WFPs) have been used for human nutrition since ancient times. The bio-cultural heritage of wild food plants, neglected and underutilized species (NUS) has been the subject of numerous studies around the world over the past decade, given that diet diversification and traditional ingredients are regarded as key issues in combating malnutrition and famine [6–9]. Not least, recognizing the significance of NUS in traditional foods and cultures helps empower indigenous groups and reaffirm their identity [10]. Furthermore, focusing on traditional plant foraging is critical in many remote areas of the world for a better understanding of its role in informing food system sustainability, providing potential health benefits, maintaining key elements embedded in the local food heritage, as well as to fostering the revitalization, reinvention and valorization of local gastronomies [11–14].

A previous study from a few Eurasian regions highlighted the importance of WFP species as a part of the traditional cuisine [15]. The role of wild plants becomes crucial when they contribute to the survival of local people in times of food shortage and unaffordability [16,17]. A recent study showed that the reliance of local people on wild plants as a source of food had increased during times of conflict [18]. Hence, the main aim of this study is to identify those culturally salient wild food plants that could become novel crops and therefore help to cope with food insecurity in specific Eurasian food-insecure regions that have recently been affected by conflicts.

2. Materials and Methods

2.1. Background for the Case Study Areas

We chose five case studies (Afghanistan, Kurdistan region of Iraq, Pakistan, Syria, and Ukraine) representing five different areas and ecosystems (Central Asia, Middle East, South Asia, Mediterranean, and Eastern Europe), respectively. The climate in Afghanistan and Pakistan is continental with low annual precipitation and long dry seasons, which lead to a short period of availability of wild food plants [19,20]. To some extent, the climate in the Kurdistan region of Iraq is similar to the one in Central Asia. On the other hand, the Mediterranean ecosystem in coastal Syria, with around 1000 mm of rainfall annually, leads to a longer period of availability of wild food plants [18]. In Ukraine, wild foraging is a forest-based practice [21]; this makes wild foraging dependent mainly on the collection of wild fruits, which appear mainly during the summer and early autumn. The five selected study areas experienced similar problems arising from conflicts and security unrest. These new realities were reflected in all life aspects including agricultural production, access to food, and local communities' perceptions towards the surrounding environment.

2.1.1. Afghanistan

Afghanistan has a distinct flora with about 5000 recognized species, 25–30% of which are endemic to the country [22]. Over four decades of continuous war, overgrazing and climate change have severely damaged the ecosystem and disturbed the human art of dependence on the environment and its biodiversity [19]. In Afghanistan, rural communities continue to gather wild food plants as a key component of their daily diet for several months of the year. Women are the main traditional ecological knowledge (TEK) holders and foragers of widely available wild plants and weeds, gathered by using just their hands from anthropogenic areas close to cultivated plots. Young boys and girls, along with senior women and men, climb mountains to collect available wild food plants and sell them cheaply in local markets to shopkeepers. Most of the wild plants gathered from mountains and weeds collected from areas close to agricultural plots are perceived as medicinal and combating hunger and malnutrition in times of famine.

2.1.2. Kurdistan Region of Iraq

Iraqi Kurdistan is a crucial area in the world in the history of agriculture. It hosts in fact some of the most important Neolithic settlements in Eurasia, such as Charmo, dating back to about 7000 BC [23,24]. Even though there is not yet a complete flora of Iraq including Kurdistan, it has been reported that in Kurdistan Region there are more than 3300 species [25]. A large proportion of these plants are located in the mountainous areas of the Kurdistan Region [26].

Iraqi Kurdistan has been the arena of some studies on WFPs in recent years, since Kurdish foraging practices are considered to be among the most vibrant and resilient ones in the world [21,23,24]. The turbulences of the past decades following the fall of Saddam Hussain's regime and especially those linked to areas which have been recently controlled by the so-called "the Islamic State" have increased the uncertainty in terms of food security. WFPs and especially wild vegetables during the spring are prominent in the local Kurdish foodscapes, making Kurdistan possibly the most iconic hotspot of wild vegetable diversity we have in Eurasia. However, the custom of knowing and gathering wild edible plants is mainly retained by elderly people, while young community members have often lost this heritage, despite local markets still offering a remarkable spectrum of diverse species to urban populations.

2.1.3. Pakistan

In Pakistan, recent studies have shown that the consumption of WFPs has remarkably decreased in the North Western belt of the country. The foraging practices are almost on the verge of extinction in many local communities. Most often, people rely on cultivated vegetables and commoditized food ingredients; therefore, ethnobotanical knowledge of WFPs is gradually decreasing and is quite threatened in certain localities. In the past, many plants have been part of traditional food systems across different cultural groups [20,27–31].

For the last two to three decades, remarkable social change has significantly impacted local foraging practices. Food mobility has altered the traditional food systems even in highly remote rural and mountain areas [30]. There are myriad of factors that may have restructured the local food systems and the foraging of WFPs. For instance, mass migration from rural and mountain areas towards cities is one of the prominent factors which has led people to rely on commercial food ingredients. Military conflicts that have been taking place in many parts across the tribal belt in western Pakistan have also indirectly affected land use management and access to foraging patches and horticultural practices. Especially in South Waziristan, the fragile security situation has led many people to abandon the local horticultural practices, while many households fled their villages and now they only come back to their villages for a very short time in summer.

2.1.4. Syria

Wild plants form an essential part of the traditional Mediterranean food system [7], and specifically wild vegetables are considered the "hidden Mediterranean diet" [32]. In coastal Syria, as a part of the Eastern Mediterranean basin, cooked vegetables and salads made from wild greens have been particularly important as local traditional foods since ancient times [33]. A recently published study from the coastal region of Syria documented 75 plant species used for food and beverage preparation [18]. Another study highlighted how wild plants contribute to the beverage culture among Syrian residents and diaspora [34]. The conflict in Syria started almost 12 years ago, and so far, it has caused mass food insecurity in the country where 60% (more than 12.4 million people) suffer from insecurity in securing their daily meals [35]. Sulaiman et al. stated that two-thirds of the study respondents reported an increase in their use of wild food plants during the conflict compared to the pre-conflict era [18]. This clearly shows how significant is the role of wild plants in the resilience of people under conflict conditions. The sustainability of wild plant use is another crucial aspect that has been affected in recent years; local people reported that the abundance of several species has significantly decreased [15].

2.1.5. Ukraine

Foraging for wild plants (especially berries) and mushrooms is an important practice for millions of people living in Ukraine. This activity is perceived as a way to perpetuate the connectedness to the surrounding landscape. Nevertheless, the current war in Ukraine has resulted in an abrupt change in the people-nature relationship [36]. This mainly disconnects Ukrainians from their landscapes by preventing them from moving freely and by pushing them into a massive forced emigration and internal relocation [36,37]. This change is represented by the displacement of local people to safe areas (which are mainly rural areas) and avoiding the movement in open natural spaces in fear of mines and military operations. These changes have been reflected in a dramatic decrease in foraging activity and wild fruit collection. However, due to economic uncertainties such as increasing prices for food and unemployment, the demand for those products is increasing.

Due to the war, 90% of Ukrainians could face extreme economic vulnerability and poverty [38], possibly resulting in an increase in the demand for wild foods.

2.2. *Data Collection and Data Analysis*

Our present study is based on a qualitative comparative case method [39]. We mainly rely on unpublished qualitative data collected from the five countries corresponding to our long-term research areas (Table 1). The selected case studies are well-established field sites of the authors, and we collected data over time through participant observation and semi-structured interviews with local people (mainly elderly people). We selected five species from each study area. The selection criteria of the species were similar among all study areas. The selected species fulfil one or more of the following criteria: it has a high value in local food culture, is used quite often by the local communities, and/or has promising economic value represented by market demand. Study participants were asked about wild food plants, their uses, mode of preparation and consumption. Visits to local markets were conducted to record the availability and prices of the reported species. Local informants were asked to estimate the species abundance in the wild. We recorded the botanical characteristics of the species as well as their local cultural value. Informants were asked whether there were any attempts to cultivate the reported species. We followed the Code of Ethics of the International Society of Ethnobiology [40]. Proper botanical identification was conducted based on the national floras and plant specimens deposited in herbaria in the study areas (Herbarium of the Department of Botany at the University of Swat in Pakistan; Estonian University of Life Sciences herbaria; the Herbarium of the American University of Beirut; Rozthochya Nature Reserve) which store voucher specimens from Pakistan, Kurdistan, Syria, and Ukraine.

Table 1. The case studies conducted on wild food plants in five Eurasian countries.

Country	Region	Interviewers
Afghanistan	Diverse areas across the country	A.K.M., A.F.
Kurdistan region of Iraq	Autonomous Kurdish Region	A.P., R.S., H.M.
Pakistan	North Western regions	M.A.A.
Syria	Coastal region	N.Su.
Ukraine	Western *oblasts*	N.St., G.M.

The collected data was mainly qualitatively analyzed, with an intention to deeply understand the importance of highlighted species and the characteristics that may promote their domestication. The analysis exhibited the ethnobotanical uses of the reported species, as well as their socio-cultural significance. To determine the market value of some species, we drew a comparison between the wild species' prices and other cultivated crop prices, while highlighting the average wages in the study areas. We assessed the environmental sustainability aspect by analyzing the estimated abundance of the species as well as observing the foraging patterns in the study areas. In addition, we reviewed the literature on the nutritional value of some species. Our analysis and discussion demonstrated the possibilities for and obstacles to cultivating the selected species.

3. Results and Discussion

3.1. Ecology, Diversity, and Nutritional Value of the Reported Species

The reported species belong to 20 genera and 16 families (Table 2). Those with herbaceous growth habits form 68% of the reported species, compared to 24% of shrubs and 8% of trees. The diversity of the reported species can be attributed to the ecosystem differences of the reported species. However, the results show that there is an overlap in the used species between the study regions. Species such as *Gundelia tournefortii* and *Chenopodium album* are used in different ecosystems and food cultures, which demonstrate the unique gastronomic characteristics of these species. The shrub-dominant habit of species growth in Ukraine explains the forest-oriented foraging pattern in the region; on the other hand, we observed the herb-dominant habit of species growth in Central Asia and the Middle East which relates to the pastoralism-based human ecological interactions.

Many of the reported species are leafy vegetables which are an important source of vitamins and minerals; this includes species such as *Malva sylvestris* (rich in Ca, Mg, and K), and *Gundelia tournefortii* (Ca, K, Na, P, Fe, Mg, and vitamin E) [41,42]. The zinc-rich species such as *Allium ampeloprasum* and *Gundelia tournefortii* could be crucial for the immune system [43]. *Capparis spinosa* possesses various biological activities, including antioxidant, antidiabetic, and anticancer, in addition to being a rich source of crude protein [44]. *Chenopodium album* leaves are rich in proteins (4.2%) with a high proportion of essential amino acids, as well as in fibers and vitamin C [45].

3.2. Mode of Preparation and Consumption

The reported species from Afghanistan were prepared in several ways. *Allium rosenbachianum* (locally called *Kheze*) is mainly gathered from the mountains surrounding the capital. Leaves of *A. rosenbachianum* are freshly gathered, washed and then boiled in water. Subsequently, this water is discarded, and leaves are fried with onion and tomatoes. This dish is eaten with mud oven-made bread *tandori naan*. A semi-similar way of preparation is followed with *Capparis spinosa* (locally called *Kevera*) where the fresh flowering buds are gathered, dried and fried with onion, tomatoes along with eggs for better taste (Figure 1). This dish is also eaten with traditionally prepared bread *tandori naan*. On the other hand, local people consume *Eremurus afghanicus* (locally known as *Siech*) in a variety of traditional ways such as with rice, or as a soup mixed with beans, mung bean and pea; while in the northern part of Afghanistan, the plant is chopped and used as a filling for the national Afghan dish *bolani* (a flat-bread, normally baked with a vegetable filling). On the other hand, the underground stem part of *Rheum spiciforme* is consumed fresh, while the fruits of *Quercus* spp. are processed into flour and used for the preparation of acorn bread (*pragi* or *nan-e-bloot*).

In the Kurdistan region of Iraq, as in several other Middle Eastern areas, *Gundelia tournefortii* (Kurdish: *Kingr*) is one of the most common foods in the spring (Figure 2). The internal lower aerial parts and sometimes roots are taken out from the soil and, after removing the thorny parts, are boiled for one to two hours and successively fried in oil with salt, onions, and garlic; another way of preparing the plant is in *Kingr kabab*, a traditional Kurdish food where eggs and flour are mixed with the vegetable and fried in oil. The way of preparing *Arum* spp. leaves (Kurdish: *Kardw*) is similar to *Kingr*, but one main difference is that *Arum* species, because of their toxicity, need to be first detoxified with water and sumac (*Rhus coriaria* L.). *Rheum ribes* stalks (Kurdish: *Rewas*) and *Dichoropetalum aromaticum* (Kurdish: *Baraza*) aerial parts are considered crucial snack foods consumed within the domestic arena with family members and friends, and which have a dense social and nutritional meaning in the local culture (*Baraza* exclusively in the East of Iraqi Kurdistan). *Quercus aegilops* (Kurdish: *Barw*) fruits are instead harvested and roasted all over the region. Oak trees have a crucial cultural meaning in the Kurdish culture: they represented the main staple in times of famine, also via acorn bread-like preparations, while they sometimes provide in early summer the proverbial Kurdish manna, which is considered a delicacy and medicine in Kurdistan [46]. Oak trees have also recently been the object of a specific campaign organized by the local authorities, which aimed to cultivate one million *Quercus* trees; this is the biggest campaign of reforestation ever started in the Kurdistan region.

Table 2. Top culturally salient WFPs holding potential for domestication in five conflict-affected contexts across Asia and Eastern Europe.

Case Study	Species Name, Family	Part Used	Growth Habit	Estimated Environmental Sustainability	Occurrence in Local Markets	Attempts for Cultivation
Afghanistan	*Allium rosenbachianum* Regel, Amaryllidaceae	Leaves	Herb	High	Widespread	No
	Capparis spinosa L., Capparaceae	Flowering buds	Shrub	Moderate	Widespread	No
	Eremurus afghanicus Gilli, Asphodelaceae	Leaves	Herb	High	Widespread	No
	Quercus spp., Fagaceae	Fruits	Tree	High	Widespread	No
	Rheum spiciforme Royle and other *Rheum* spp., Polygonaceae	Stalks	Herb	High	Widespread	No
Kurdistan region of Iraq	*Arum* spp., Araceae	Aerial parts	Herb	High	Widespread	No
	Dichoropetalum aromaticum (Rech.f.) Pimenov & Kljuykov, Apiaceae	Aerial parts	Herb	Moderate	Widespread	No
	Gundelia tournefortii L., Asteraceae	Young leaves and roots	Herb	High	Widespread	No
	Quercus aegilops L., Fagaceae	Fruits	Tree	High	Widespread	No
	Rheum ribes L., Polygonaceae	Stalks	Herb	High	Widespread	No
Pakistan	*Amaranthus viridis* All., Amaranthaceae	Aerial part	Herb	Critically low	Almost absent	No
	Chenopodium album L., Amaranthaceae	Aerial part	Herb	Moderate	Rare	No
	Lepidium draba L., Brassicaceae	Aerial part	Herb	High	Widespread	No
	Malcolmia africana (L.) W.T.Aiton, Brassicaceae	Aerial part	Herb	High	Widespread	No
	Rumex spp., Polygonaceae	Aerial part	Herb	High	Widespread	No
Syria	*Allium ampeloprasum* L., Amaryllidaceae	Young aerial part and bulb	Herb	Moderate to low	Rare	No
	Anchusa strigosa Banks and Sol., Boraginaceae	Young aerial part	Herb	Low	Fair	No
	Crataegus azarolus L., Rosaceae	Fruits	Shrub	Low	Rare	Yes
	Gundelia tournefortii L., Asteraceae	Leaves midrib and underground stem	Herb	Critically low	Almost absent	Yes
	Malva sylvestris L., Malvaceae	Young aerial part	Herb	High	Widespread	No
Ukraine	*Chenopodium album* L., Amaranthaceae	Young aerial part	Herb	High	Widespread	No
	Corylus avellana L., Betulaceae	Kernels	Shrub	High	Widespread	Yes
	Hippophae rhamnoides L., Elaeagnaceae	Fruits	Shrub	High	Widespread	Yes
	Rubus idaeus L., Rosaceae	Fruits	Shrub	High	Widespread	Yes
	Vaccinium myrtillus L., Ericaceae	Fruits	Shrub	High	Widespread	No

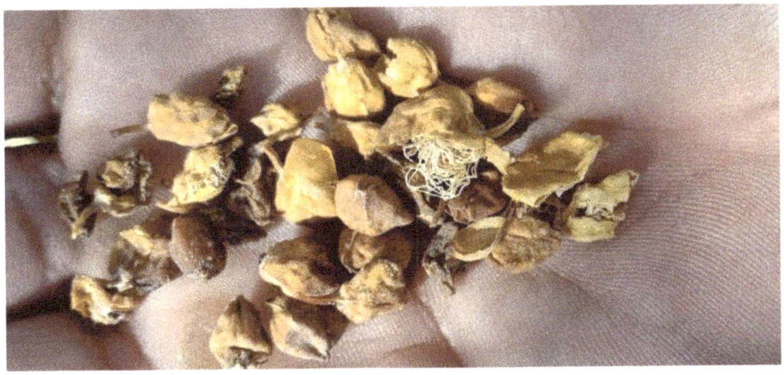

Figure 1. Dried flower buds of *Capparis spinosa* (Photo: Manduzai A.K.).

Figure 2. Wild food plants (*Arum* spp. in the middle upper part of the picture and *Gundelia tournefortii* in the bottom) sold in a local market in Kurdistan (Photo: Ahmed H.M.).

In NW Pakistan, all of the reported species are consumed as cooked vegetables. However, some species such as *Lepidium draba* are also prepared as a salad or consumed as a raw snack (Figure 3). Some of the reported species are highly used in other parts of Pakistan as several studies from other regions reported the use of *Amaranthus* spp. [20,27,28,31]. Similarly, *Rumex* spp. is highly consumed and forms an essential part of cooked wild vegetables.

Figure 3. A local vendor selling WFPs (i.e., Eremurus *stenophyllus, Lepidium draba* subsp. *chalepense, Rumex dentatus*) in Quetta Bazar in Pakistan (Photo: Aziz M.A.).

Out of the five reported WFPs in Syria, four species are prepared by steaming the young aerial part with some food compounds. "*Sleeq*" is a popular wild plant-based dish where several wild leafy vegetables are steamed together with onion and olive oil. However, each of the reported species can be prepared in several ways. For instance, *A. ampeloprasum* can be consumed fresh as an appetizer, fried with eggs, or steamed with other wild leafy vegetables (Figure 4). *G. tournefortii* is a highly preferred species for its unique taste; it is usually steamed with chickpea or minced meat. The species *M. sylvestris* (locally called *Khebbaizeh*) is steamed with onion, olive oil, and bulgur. On the other hand, *C. azarolus* is consumed as a snack, especially when local people walk in the wild.

Fruits and nuts are the consumed part of most of the reported species in Ukraine; these can be explained by the forest-based forage ecosystem in the country. The nuts of *Corylus avellana* are mainly used in baked cookies and pies such as the traditional cookies of "Swallow's Nest" as well as the Christmas ritual dish "*Kutya*" (Figure 5). The nuts are also used as a snack and as an addition to some desserts. The fruits of *Vaccinium myrtillus* are consumed in various ways. The fruits can be preserved by keeping them with sugar in cold places over the winter. They are also used in preparing different jams and tinctures. Fresh fruits of *V. myrtillus* are used for traditional *varenyky* (a kind of tortellini with berries). The fruits can also be boiled and served with sour cream. Another mode of preparation is by using fresh fruits as ingredients in pies. Blueberries (*V. myrtillus*) are also used to make juices, fruit drinks, extracts, syrups, compotes and marmalades. They are also preserved dried to be used as tea or in pies and other preparations in wintertime [21]. Raspberry (*Rubus idaeus*) is another species whose fruits are consumed. It can be eaten fresh or used to make jam, jelly, marmalade, pastille, and juices. Raspberry wines, tinctures, and liqueurs are characterized by high-taste properties and exceptional aroma. The fruits can also be dried or preserved by grinding them with sugar. On the other hand, the aerial part of *Chenopodium album* is used (locally called *Loboda*). The species was used intensively in the past when there was little to eat in the spring. *C. album* is used in soups (e.g., green borscht), stews and fried vegetables. It could be used fresh or salted. The young aerial part is washed well before use by soaking it in a salty liquid. The plants can be preserved by spreading or hanging them in bundles in the open air and then storage in glass jars or wooden boxes lined with paper.

Figure 4. Allium *ampeloprasum*; (Photo: Sulaiman N).

Figure 5. Hazelnut (*Corylus avellana*); (Photo: Stryamets S).

Table 2 clearly shows that the dominant modes of preparation and consumption among the reported species are the cooked green part of the species, and the snack consumption of the raw fruits. However, some species could provide other uses such as grains in *Chenopodium album*, and *Amaranthus viridis*.

3.3. Economic Value of the Reported Species

All the reported species were found to be sold in local markets. More than three-quarters (76%) of the species are reported as widespread in the local markets (Table 2). In Afghanistan, Pakistan, and Kurdistan, young boys and girls, along with senior women and men, climb mountains to collect available wild food plants and sell them cheaply in local markets to shopkeepers. On the other hand, rural women in Syria collect wild plants from the surrounding communal lands and orchards and sell them to local shops or directly to customers in local markets. In Ukraine, both women and men collect wild species from the nearby forests and sell them to local markets. Some of our respondents in Syria reported that income generated from selling wild plants reaches up to 20% of their annual income. However, security concerns may limit these practices quite often as all our case study areas have experienced conflict and security unrest.

The prices of most of the wild plants in local markets are relatively lower than the cultivated species. This drives many people, especially those who live in the cities and town centers and cannot access the wild, to buy WFPs because they are more affordable in some study areas. However, some wild species are more expensive than the other cultivated species, as those wild plants are perceived as organic and healthier products. In addition, some species become rare in nature while the demand is high; this is reflected in relatively high prices. Table 3 shows the price of several species compared to the approximate price of other cultivated vegetables and fruits, and to the mean monthly salary [47].

Table 3. Comparison of the prices of the most iconic wild food plants and of cultivated species in the five selected case studies.

Case Study	Species Name	Price per Unit	Price of 1 kg Tomato	Price of 1 kg of Apple	Mean Monthly Salary (USD)
Afghanistan	*Allium rosenbachianum* *Capparis spinosa*	2 USD/kg 5 USD/kg	0.5 USD/kg	0.7 USD/kg	50
Iraq	*Arum* spp., *Rheum ribes*	3 USD/kg 2.05 USD/kg	0.68 USD/kg	0.7 USD/kg	1000
Pakistan	*Amaranthus viridis* *Chenopodium album*	1 USD/kg 0.9 USD/kg	2.2 USD/kg	2 USD/kg	365
Syria	*Anchusa strigosa* *Malva sylvestris*	0.3 USD/kg 0.2 USD/kg	0.4 USD/kg	0.7 USD/kg	30
Ukraine	*Corylus avellana* *Vaccinium myrtillus*	5 USD/kg of unpeeled nuts 5 USD/kg	1.7 USD/ kg	0.35 USD/kg	625

3.4. Sustainability Status of the Reported Species

Many of the reported species have been listed in the Red List Book such as *Rubus idaeus*, *Capparis spinosa*, *Vaccinium myrtillus*, *Corylus avellana*, *Crataegus azarolus*, and *Dichoropetalum aromaticum* [48]. An alarm status has been reported for some of these species such as *Crataegus azarolus*, and some other unlisted species such as *Amaranthus viridis*, *Allium ampeloprasum*, and *Gundelia tournefortii*, for which locals reported a noticeable decrease in the species' availability. This can be attributed to the unsustainable manner of harvesting in the case of *Allium ampeloprasum* as the whole plant is pulled with its bulb before reaching the flowering stage, while the case is different in *Crataegus azarolus* as the shrub is being largely cut to be used as firewood due to the fuel deficiency during the conflict in Syria. On the other hand, species such as *Dichoropetalum aromaticum* and *Capparis spinosa* are gathered in a sustainable way (only aerial parts are collected), but the common over-foraging of these species, due to their cultural and economic value, may raise a sustainability concern on the long term. *Vaccinium myrtillus* is largely available in (Western) Ukraine, yet its productivity depends on the year and on the intensity of the previous harvest. In some areas of Ukraine, we observed an over-harvesting status of berries due to the crucial importance of this product for local economies of some Carpathian villages [15]. *Corylus avellana* is widely used as a building material for fences, decorations, and garden braces. In the steppes of Ukraine, *Corylus avellana* is used as firewood, and this may explain its unstable abundance status.

3.5. A Shift from Wild to Cultivated: Possibilities and Obstacles

All the reported species have high cultural importance as they form part of the traditional cuisine and are used on specific occasions (e.g., *C. avellana* is used in preparing a ritual Christmas dish in Ukraine). Drawing from our direct observations in the field, the reported species became crucial for local people in recent years after the study areas experienced tough economic circumstances linked to wars and socio-political turbulence. In a few cases, these species served as main food sources for many families. Economically, since all the highlighted species in Table 2 are sold in local markets, this suggests that these species have a considerable market demand and a promising economic value. Moreover, the potential new crops are supposed to be more eco-friendly as they are well adapted to local environmental conditions and thus can withstand biotic and abiotic stresses. Agricultural inputs (e.g., fertilizers and pesticides) will be less needed with the new potential crops as they have higher genome diversity and are more resistant to diseases and pests, as well as more adapted to different soil conditions. In addition, plant propagation materials (e.g., seeds, cuttings, division, grafting) are widely available in the wild. Thus, farmers will be more independent in their farming from commercial propagation materials. We observed that most of our study participants have either a piece of land or a home garden where they are able to practice horticulture, especially growing different vegetables. Some of our respondents reported difficulties experienced in foraging in some periods of the recent past, due to security concerns. Therefore, the cultivation of the foraged species could possibly provide rural households with more diverse and nutritious food resources without the need to forage in remote areas or to put themselves in unsafe/risky situations. For all the aforementioned reasons, the possibility of cultivating these species represents a serious option to be considered.

However, there are several obstacles to the exploration of this option. This is mainly related to the initiatives that should be taken by community leaders and local agricultural institutions. Moreover, the decreased abundance of these plants in the wild is another critical issue; therefore, the start of such a project for taming them could also serve as a bio-conservation measure. Agronomic feasibility is another crucial factor that needs to be studied before initiating cultivation programs in order to examine the yield and the potential pests and diseases that could challenge the cultivation. Molina et al. studied the agronomic feasibility of *Cichorium intybus* in the Mediterranean conditions of Spain and found that it has a higher yield than in its natural habitat (7016 kg/ha), whereas the yield of *Rumex pulcher* was 4923 kg/ha [49]. The same study highlighted the yield of some edible species in the wild such as the bulbs of *Allium ampeloprasum* and the fruits of *Crataegus monogyna* were around 250 kg/ha and 500 kg/ha, respectively. Another study reported a similar production yield of *Allium ampeloprasum*, and found the yield was positively correlated with monthly precipitation at harvest time [50]. However, cultivation experiments have to take into consideration that these values differ between ecoregions based on several factors such as soil, precipitation, temperature among the seasons, and plant varieties. Domestication of wild species through traditional agronomic methods is supposed to be ethically and socially feasible as it does not contain any genetic-editing techniques as in some other domestication processes [51]. In addition to species yield, taste quality is among the major traits that have to be considered when domestication programs are applied, as the gastronomic characteristics of the species will crucially affect their market demand.

Another crucial aspect that highlights the importance of the listed species is that some of them are considered to be wild crop relatives such as *Allium ampeloprasum* as a wild relative to the cultivated garlic and leek (*Allium sativum* and *Allium porrum*), and *Corylus avellana* as a relative to *Corylus maxima* [52]. Thus, these wild species could play an important role in breeding such crops and conserving their diverse genetic materials.

3.6. Domestication as a Mean to Safeguard and Reinforce the Food Heritage of Local Communities: Potential Risks and Unintended Side Effects

All the reported species are embedded into the food culture and gastronomy of the studied local communities. However, the dynamics at play have hindered the use of some of these species, given the increasing detachment of local dwellers from the foodscapes where these species have been traditionally managed and harvested due to security issues related to the conflictual situations our case studies are facing. This could affect the persistence of the traditional gastronomic knowledge attached to these species. In this sense, the transformation of some of the identified species into novel cultivated crops could positively impact the maintenance and reinforcement of some key traits of the traditional gastronomic culture and associated heritage, also in post-conflict times.

However, this process is not free from possible frictions and side effects that could affect both the social and cultural value of these species, as well as their role in the diet of local communities. An increase in commercial value and adaptations to the market dynamics could negatively reverberate on the conservation of socio-cultural values attached to these species. Moreover, as already shown elsewhere [53–55], the valorization of wild foods and food-related resources could negatively affect the sustainability of the species and the ecosystem. While the issue could be possibly overcome through the cultivation of some of the identified species, the transformation of foods traditionally linked to subsistence and local diets into marketable products could run the risk of fostering their commodification (i.e., cultural and heritage commodification), eroding the values linked to specific social and cultural practices, shifting their role from subsistence foods to cash crops, and triggering the co-optation of products and associated knowledge by extra-local economic actors [56].

In order to potentially limit these side effects and create positive externalities in terms of the food security and traditional food heritage of local communities, a balance between the promotion of these foods for household consumption and as a source of income should be found. This would, in turn, require further participatory research aimed at exploring the possible scenarios related to the valorization processes (i.e., increase in the social and economic value and desirability of a product), focusing on potential changes in the roles that these species have traditionally played in the livelihoods and culture of local communities.

4. Conclusions

The present paper contextualizes and compares five Eurasian case studies that underwent similar socioeconomic realities arising from conflict situations and security unrest. We highlighted a total of 20 taxa with high potential to become novel cultivated plants in the present and/or in the post-conflict time. The reported species have high cultural importance for the local communities and have promising economic value. The suggested wild species for cultivation could improve the nutritional status of local communities and their security (as they can cultivate them close to their houses, avoiding the risk of foraging in areas distant from inhabited areas), and they could improve the economic situation for the household as they may contribute to income generation. The cultivation of the species will also significantly contribute to the conservation of vulnerable species, especially those highlighted by the Red List Book. The reported species in the present study may also be suitable for cultivation in the neighboring regions of our study areas. In addition, this study may contribute to drawing the attention of the scientific community to the importance of wild plants as possible novel crops, especially under the circumstances of climate change and biodiversity loss. Future studies should focus on the agronomic feasibility of the highlighted species within their regional ecosystems. Afterwards, local agrarian, NGO and other institutions could consider these local plants for widespread cultivation as a way to reduce food insecurity in conflict areas.

Author Contributions: Conceptualization and methodology: A.P., N.S. (Naji Sulaiman), G.M. and D.M.Z.; investigation: N.S. (Naji Sulaiman), M.A.A., N.S. (Natalie Stryamets), G.M., H.M.A., A.K.M., A.F., R.S. and A.P.; data curation: N.S. (Naji Sulaiman); writing—original draft preparation: N.S. (Naji Sulaiman); writing—review and editing: A.P., Z.P., D.M.Z., G.M., N.S. (Natalie Stryamets), M.A.A., H.M.A., A.A.S., A.F., R.S. and N.S. (Naji Sulaiman). All authors have read and agreed to the published version of the manuscript.

Funding: The fieldwork in Syria was supported by the Internal Grant Agency of the Faculty of Tropical AgriSciences, Czech University of Life Sciences, Prague, Czech Republic (IGA FTZ, Project No. 20223104); while the fieldwork in Ukraine was supported by the European Research Council (ERC) under the European Union's Horizon 2020 research and innovation programme (grant agreement No 714874); the authors are also thankful to the University of Pollenzo, Italy, for having funded the field research in Iraqi Kurdistan from 2016 to 2019.

Data Availability Statement: Not applicable.

Acknowledgments: The authors would like to thank local people for their time and knowledge, especially in the hard conditions they experience.

Conflicts of Interest: The authors declare no conflict of interest.

References

1. Von Wettberg, E.; Davis, T.M.; Smýkal, P. Editorial: Wild Plants as Source of New Crops. *Front. Plant Sci.* **2020**, *11*, 591554. [CrossRef] [PubMed]
2. Willis, K.J. *State of the World's Plants*; Kew, Royal Botanic Gardens: London, UK, 2017. Available online: https://library.oapen.org/handle/20.500.12657/49594 (accessed on 26 December 2022).
3. FAO. Harvesting Nature's Diversity. 2004. Available online: http://www.fao.org/DOCREP/004/V1430E/V1430E04.htm (accessed on 25 December 2022).
4. Hajjar, R.; Hodgkin, T. The use of wild relatives in crop improvement: A survey of developments over the last 20 years. *Euphytica* **2007**, *156*, 1–13. [CrossRef]
5. Michon, G.; De Foresta, H. Agroforests: Pre-domestication of forest trees or true domestication of forest ecosystems? *Neth. J. Agric. Sci.* **1997**, *45*, 451–462. [CrossRef]
6. Sõukand, R. Perceived reasons for changes in the use of wild food plants in Saaremaa, Estonia. *Appetite* **2016**, *107*, 231–241. [CrossRef]
7. Pieroni, A.; Sulaiman, N.; Sõukand, R. Chorta (Wild Greens) in Central Crete: The Bio-Cultural Heritage of a Hidden and Resilient Ingredient of the Mediterranean Diet. *Biology* **2022**, *11*, 673. [CrossRef] [PubMed]
8. Ulian, T.; Diazgranados, M.; Pironon, S.; Padulosi, S.; Liu, U.; Davies, L.; Howes, M.-J.; Borrell, J.; Ondo, I.; Pérez-Escobar, O.; et al. Unlocking Plant Resources to Support Food Security and Promote Sustainable Agriculture. *Plants People Planet* **2020**, *2020*, 421–445. [CrossRef]
9. Vorstenbosch, T.; de Zwarte, I.; Duistermaat, L.; van Andel, T. Famine food of vegetal origin consumed in the Netherlands during World War II. *J. Ethnobiol. Ethnomed.* **2017**, *13*, 63. [CrossRef] [PubMed]
10. Padulosi, S.; King, E.I.O.; Hunter, D.; Swaminathan, M.S. *Orphan Crops for Sustainable Food and Nutrition Security: Promoting Neglected and Underutilized Species*, 1st ed.; Routledge: London, UK, 2021.
11. Padulosi, S.; Amaya, K.; Jäger, M.; Gotor, E.; Rojas, W.; Valdivia, R. A Holistic Approach to Enhance the Use of Neglected and Underutilized Species: The Case of Andean Grains in Bolivia and Peru. *Sustainability* **2014**, *6*, 1283–1312. [CrossRef]
12. Borelli, T.; Hunter, D.; Powell, B.; Ulian, T.; Mattana, E.; Termote, C.; Pawera, L.; Beltrame, D.; Penafiel, D.; Tan, A.; et al. Born to Eat Wild: An Integrated Conservation Approach to Secure Wild Food Plants for Food Security and Nutrition. *Plants* **2020**, *9*, 1299. [CrossRef]
13. Borelli, T.; Hunter, D.; Padulosi, S.; Amaya, N.; Meldrum, G.; de Oliveira Beltrame, D.M.; Samarasinghe, G.; Wasike, V.W.; Güner, B.; Tan, A.; et al. Local Solutions for Sustainable Food Systems: The Contribution of Orphan Crops and Wild Edible Species. *Agronomy* **2020**, *10*, 231. [CrossRef]
14. Johns, T.; Eyzaguirre, P.B. Linking Biodiversity, Diet and Health in Policy and Practice. *Proc. Nutr. Soc.* **2006**, *65*, 182–189. [CrossRef]
15. Aziz, M.A.; Mattalia, G.; Sulaiman, N.; Shah, A.A.; Polesny, Z.; Kalle, R.; Sõukand, R.; Pieroni, A. The nexus between traditional foraging and its sustainability: A qualitative assessment among a few selected Eurasian case studies. *Environ. Dev. Sustain.* **2022**, 1–26. [CrossRef]
16. Zocchi, D.M.; Bondioli, C.; Hamzeh Hosseini, S.; Miara, M.D.; Musarella, C.M.; Mohammadi, D.; Khan Manduzai, A.; Dilawer Issa, K.; Sulaiman, N.; Khatib, C.; et al. Food Security beyond Cereals: A Cross-Geographical Comparative Study on Acorn Bread Heritage in the Mediterranean and the Middle East. *Foods* **2022**, *11*, 3898. [CrossRef] [PubMed]
17. Redžić, S.; Ferrier, J. The Use of Wild Plants for Human Nutrition During a War: Eastern Bosnia (Western Balkans). In *Ethnobotany and Biocultural Diversities in the Balkans*; Pieroni, A., Quave, C.L., Eds.; Springer: New York, NY, USA, 2014; pp. 149–182.

18. Sulaiman, N.; Pieroni, A.; Sõukand, R.; Polesny, Z. Food Behavior in Emergency Time: Wild Plant Use for Human Nutrition during the Conflict in Syria. *Foods* **2022**, *11*, 177. [CrossRef] [PubMed]
19. Yousufi, A. Horticulture in Afghanistan: Challenges and Opportunities. *J. Dev. Sustain. Agric.* **2016**, *11*, 36.
20. Aziz, M.A.; Abbasi, A.M.; Ullah, Z.; Pieroni, A. Shared but threatened: The heritage of wild food plant gathering among different linguistic and religious groups in the Ishkoman and Yasin Valleys, North Pakistan. *Foods* **2020**, *9*, 601. [CrossRef] [PubMed]
21. Pieroni, A.; Sõukand, R. Forest as Stronghold of Local Ecological Practice: Currently Used Wild Food Plants in Polesia, Northern Ukraine. *Econ. Bot.* **2018**, *72*, 311–331. [CrossRef]
22. Breckle, S.W.; Rafiqpoor, M.D. The Hindu Kush/Afghanistan. In *Plant Biogeography and Vegetation of High Mountains of Central and South-West Asia*; Noroozi, J., Ed.; Springer: Cham, Switzerland, 2020; Volume 17. [CrossRef]
23. Pieroni, A.; Ahmed, H.M.; Zahir, H. The spring has arrived: Traditional wild vegetables gathered by Yarsanis (Ahl-e Haqq) and Sunni Muslims in Western Hawraman, SE Kurdistan (Iraq). *Acta Soc. Bot. Pol.* **2017**, *86*. [CrossRef]
24. Tsuneki, A.; Rasheed, K.; Watanabe, N.; Anma, R.; Tatsumi, Y.; Minami, M. Landscape and early farming at Neolithic sites in Slemani, Iraqi Kurdistan: A case study of Jarmo and Qalat Said Ahmadan. *Paléorient Rev. Pluridiscip. Préhistoire Protohist. L'asie Sud-Ouest L'asie Cent.* **2019**, *45*, 33–51. [CrossRef]
25. Ghazanfar, S.A.; McDaniel, T. Floras of the Middle East: A quantitative analysis and biogeography of the flora of Iraq. *Edinb. J. Bot.* **2016**, *73*, 1–24. [CrossRef]
26. Ahmed, H.M. Traditional uses of Kurdish medicinal plant *Pistacia atlantica* subsp. *kurdica* Zohary in Ranya, Southern Kurdistan. *Genet. Resour. Crop Evol.* **2017**, *64*, 1473–1484. [CrossRef]
27. Aziz, M.A.; Ullah, Z.; Pieroni, A. Wild food plant gathering among Kalasha, Yidgha, Nuristani and Khowar speakers in Chitral, NW Pakistan. *Sustainability* **2020**, *12*, 9176. [CrossRef]
28. Aziz, M.A.; Ullah, Z.; Adnan, M.; Sõukand, R.; Pieroni, A. The Fading Wild Plant Food–Medicines in Upper Chitral, NW Pakistan. *Foods* **2021**, *10*, 2494. [CrossRef] [PubMed]
29. Aziz, M.A.; Ullah, Z.; Al-Fatimi, M.; De Chiara, M.; Sõukand, R.; Pieroni, A. On the Trail of an Ancient Middle Eastern Ethnobotany: Traditional Wild Food Plants Gathered by Ormuri Speakers in Kaniguram, NW Pakistan. *Biology* **2021**, *10*, 302. [CrossRef]
30. Aziz, M.A.; Ullah, Z.; Adnan., M.; Sõukand, R.; Pieroni, A. Plant Use Adaptation in Pamir: Sarikoli Foraging in the Wakhan Area, Northern Pakistan. *Biology* **2022**, *11*, 1543. [CrossRef]
31. Aziz, M.A.; Abbasi, A.M.; Saeed, S.; Ahmed, A.; Pieroni, A. The Inextricable Link between Ecology and Taste: Traditional Plant Foraging in NW Balochistan, Pakistan. *Econ. Bot.* **2022**, *76*, 34–59. [CrossRef]
32. Biscotti, N.; Pieroni, A. The hidden Mediterranean diet: Wild greens traditionally gathered and consumed in the Gargano area, Apulia, SE Italy. *Acta Soc. Bot. Pol.* **2015**, *84*, 327–338. [CrossRef]
33. Nebel, S.; Pieroni, A.; Heinrich, M. Ta chorta: Wild edible greens used in the Graecanic area in Calabria, southern Italy. *Appetite* **2006**, *47*, 333–342. [CrossRef] [PubMed]
34. Sulaiman, N.; Pieroni, A.; Sõukand, R.; Whitney, C.; Polesny, Z. Socio-Cultural Significance of Yerba Maté among Syrian Residents and Diaspora. *Econ. Bot.* **2021**, *75*, 97–111. [CrossRef]
35. WFP. Situation Report. 2021. Available online: https://api.godocs.wfp.org/api/documents/32d06a6a825341b08e344f35913ddfea/download/?_ga=2.214902231.1764341077.1657962784-1008178108.1657962784 (accessed on 20 October 2022).
36. Stryamets, N.; Prakofjewa, J.; Mattalia, G.; Kalle, R.; Pruse, B.; Zocchi, D.M.; Sõukand, R.; Pieroni, A.; Fontefrancesco, M.F. Why the ongoing occupation of Ukraine matters to ethnobiology. *J. Ethnobiol. Ethnomed.* **2022**, *18*, 21. [CrossRef]
37. UNHCR. Situation Report. 2022. Available online: https://data.unhcr.org/en/situations/ukraine (accessed on 26 December 2022).
38. UNDP. Every Day of Delayed Peace Will Accelerate a Freefall into Poverty for Ukraine, Warns UNDP. 2022. Avaiable online: https://www.undp.org/press-releases/every-day-delayed-peace-will-accelerate-freefall-poverty-ukraine-warns-undp (accessed on 26 December 2022).
39. Eisenhardt, K.M. Building Theories from Case Study Research. *Acad. Manag. Rev.* **1989**, *14*, 532–550. [CrossRef]
40. Code of Ethics. ISE (International Society of Ethnobiology) Web Site. 2008. Available online: https://www.ethnobiology.net/wp-content/uploads/ISE-COE_Eng_rev_24Nov08.pdf (accessed on 16 November 2022).
41. Matthäus, B.; Özcan, M.M. Chemical evaluation of flower bud and oils of tumbleweed (*Gundelia tourneforti* L.) as a new potential nutrition sources. *J. Food Biochem.* **2011**, *35*, 1257–1266. [CrossRef]
42. Hiçsönmez, Ü.; Ereeş, F.S.; Özdemir, C.; Özdemir, A.; Cam, S. Determination of major and minor elements in the *Malva sylvestris* L. from Turkey using ICP-OES techniques. *Biol. Trace Elem. Res.* **2009**, *128*, 248–257. [CrossRef]
43. García-Herrera, P.; Morales, P.; Fernández-Ruiz, V.; Sánchez-Mata, M.C.; Cámara, M.; Carvalho, A.M.; Ferreira, I.C.; Pardo-de-Santayana, M.; Molina, M.; Tardío, J. Nutrients, phytochemicals and antioxidant activity in wild populations of *Allium ampeloprasum* L., a valuable underutilized vegetable. *Food Res. Int.* **2014**, *62*, 272–279. [CrossRef]
44. Sun, Y.; Yang, T.; Wang, C. *Capparis spinosa* L. as a potential source of nutrition and its health benefits in foods: A comprehensive review of its phytochemistry, bioactivities, safety, and application. *Food Chem.* **2022**, *409*, 135258. [CrossRef] [PubMed]
45. Poonia, A.; Upadhayay, A. *Chenopodium album* Linn: Review of nutritive value and biological properties. *J. Food Sci. Technol.* **2015**, *52*, 3977–3985. [CrossRef]
46. Pieroni, A.; Zahir, H.; Amin, H.I.M.; Sõukand, R. Where tulips and crocuses are popular food snacks: Kurdish traditional foraging reveals traces of mobile pastoralism in Southern Iraqi Kurdistan. *J. Ethnobiol. Ethnomed.* **2019**, *15*, 59. [CrossRef]

47. Salary Explorer. 2022. Avalaible online: http://www.salaryexplorer.com/ (accessed on 23 December 2022).
48. IUCN Red List (International Union for Conservation of Nature). 2022. Available online: https://www.iucnredlist.org/ (accessed on 26 December 2022).
49. Molina, M.; Pardo-de-Santayana, M.; Tardío, J. Natural Production and Cultivation of Mediterranean Wild Edibles. In *Mediterranean Wild Edible Plants*; Sánchez-Mata, M., Tardío, J., Eds.; Springer: New York, NY, USA, 2016. [CrossRef]
50. Molina, M.; Tardío, J.; Aceituno-Mata, L.; Morales, R.; Reyes-García, V.; Pardo-de-Santayana, M. Weeds and food diversity: Natural yield assessment and future alternatives for traditionally consumed wild vegetables. *J. Ethnobiol.* **2014**, *34*, 44–67. [CrossRef]
51. Andersen, M.M.; Landes, X.; Xiang, W.; Anyshchenko, A.; Falhof, J.; Østerberg, J.T.; Olsen, L.I.; Edenbrandt, A.K.; Vedel, S.E.; Thorsen, B.J.; et al. Feasibility of new breeding techniques for organic farming. *Trends Plant Sci.* **2015**, *20*, 426–434. [CrossRef]
52. Crop Wild Relative Inventory. 2022. Available online: https://www.cwrdiversity.org/checklist/ (accessed on 27 December 2022).
53. De Mello, N.G.R.; Gulinck, H.; Van den Broeck, P.; Parra, C. Social-Ecological Sustainability of Non-Timber Forest Products: A Review and Theoretical Considerations for Future Research. *For. Policy Econ.* **2020**, *112*, 102109. [CrossRef]
54. Roe, D.; Booker, F.; Day, M.; Zhou, W.; Allebone-Webb, S.; Hill, N.A.O.; Kumpel, N.; Petrokofsky, G.; Redford, K.; Russell, D.; et al. Are Alternative Livelihood Projects Effective at Reducing Local Threats to Specified Elements of Biodiversity and/or Improving or Maintaining the Conservation Status of Those Elements? *Environ. Evid.* **2015**, *4*, 22. [CrossRef]
55. Kusters, K.; Achdiawan, R.; Belcher, B.; Ruiz Perez, M. Balancing Development and Conservation? An Assessment of Livelihood and Environmental Outcomes of Nontimber Forest Product Trade in Asia, Africa and Latin America. *Ecol. Soc.* **2006**, *11*. [CrossRef]
56. Zocchi, D.M.; Fontefrancesco, M.F.; Corvo, P.; Pieroni, A. Recognising, Safeguarding, and Promoting Food Heritage: Challenges and Prospects for the Future of Sustainable Food Systems. *Sustainability* **2021**, *13*, 9510. [CrossRef]

horticulturae

MDPI

Article

Preliminary Assessment of Four Wild Leafy Species to Be Used as Baby Salads

Ada Baldi *, Stefania Truschi, Piero Bruschi and Anna Lenzi

Department of Agriculture, Food, Environment and Forestry (DAGRI), University of Florence Piazzale delle Cascine 18, 50144 Firenze, Italy; stefania.truschi@unifi.it (S.T.); piero.bruschi@unifi.it (P.B.); anna.lenzi@unifi.it (A.L.)

* Correspondence: ada.baldi@unifi.it

Abstract: Wild edible leafy plants, thanks to their organoleptic characteristics and nutritional value that can make them be appreciated as salads by consumers, represent a good opportunity for growers and the fresh-cut industry, which are always looking for new crops to expand the number of products they offer. In this study, four wild species (dandelion, sorrel, wild chicory, and wild lettuce) were cultivated hydroponically up to the baby leaf stage in order to evaluate them as potential crops. At harvest, yield and antioxidant compounds, minerals, and nitrates content were assessed. The contribution to human mineral intake and the possible health risk associated with heavy metals were investigated. A characterization of the sensory profile was also carried out. Yield and chlorophylls and carotenoids content of the investigated species were comparable to those of common leafy vegetables. Variability in nitrate content was observed, with the lowest value in sorrel and the highest in dandelion. All species could contribute in Cr, Mg, and Se intake, and health risks due to heavy metals were excluded. Each species was well characterized by distinctive and peculiar sensory notes. In conclusion, the results of this preliminary study suggest that the four wild investigated species may be promising for baby leaf production.

Keywords: *Taraxacum campylodes* G.E.Haglund; *Rumex acetosa* L.; *Cichorium intybus* L.; *Lactuca serriola* L.; leafy vegetables; yield and quality; nitrate; dietary intake; health risk; sensory profile

Citation: Baldi, A.; Truschi, S.; Bruschi, P.; Lenzi, A. Preliminary Assessment of Four Wild Leafy Species to Be Used as Baby Salads. *Horticulturae* **2023**, *9*, 650. https://doi.org/10.3390/horticulturae9060650

Academic Editor: Yuyang Zhang

Received: 16 May 2023
Revised: 26 May 2023
Accepted: 29 May 2023
Published: 1 June 2023

1. Introduction

The term "baby leaf" refers to the young leaves of vegetable crops harvested up to the eighth true leaf and are mainly consumed as salads, both singularly and in multi-species mixes [1]. Although possibly commercialized as unprocessed products, baby leaves are usually minimally processed as ready-to-eat salads, a market globally valued at USD 10.78 billion in 2020 and expected to expand further in the coming years [2]. The worldwide success of this food category is to be found in the fact that it combines excellent health properties due to the high content of nutrients and antioxidant compounds, coupled with fresh consumption, with the advantage of ease of use; thus, it meets both the demand of consumers increasingly aware of the health benefits of a diet rich in fresh vegetables and the need for quick food preparation typical of modern life [3]. For the fresh-cut industry, baby leaves have the advantage of undergoing a lower degree of cutting than adult vegetables, which makes them less subject to browning of the surfaces of the cut edges and to the release of nutrients favorable to the growth of bacteria [4].

A high number of species belonging to different botanical families are grown as baby leaves, of which lettuce (*Lactuca sativa* L.) (Asteraceae), with many types, is the most important, followed by rocket (*Eruca vesicaria* ([L.] Cav.) and wild rocket (*Diplotaxis tenuifolia* ([L.] DC.) (Brassicaceae) [1,3]. The large assortment of baby leaf crops provides a wide range of shapes, colours, tastes, and textures. Nevertheless, consumers are constantly demanding diversification of products, and consequently, growers and the fresh-cut industry are always looking for new crops to expand the number of goods they offer. Ethnobotany may be a source of inspiration for this purpose, as wild edible flora is rich in leafy species that are gathered and consumed as salad [5]. In Mediterranean countries, wild greens have always been an important part of the daily diet [6]. For example, in the Tuscany region (Italy), among 357 taxa of wild food plants traditionally used in the local gastronomy, 220 are herbaceous plants utilized as leafy vegetables, and, of 17 different recipe groups, salads are the second largest category [5]. Some of these species (e.g., *Cichorium intybus* L.) are progenitors of crops but could further be exploited for agricultural purposes, as during domestication, the strong selective pressure could have led to the loss of precious alleles and characters [7]. Some others (e.g., *Taraxacum campylodes* G.E.Haglund) have been subjected to domestic or semi-amateur cultivation experiences (very small areas, local markets) but deserve to be taken into consideration for cultivation on a larger scale. Finally, other interesting wild edible plants are still completely uncultivated [8]. For example, among Asteraceae, *Helminthotheca echioides* (L.) Holub and *Hypochoeris radicata* L., present in the list of the first 30 most cited wild edible species in Tuscany according to the ethnobotanical study of Baldi et al. [5], have never been cultivated.

In general, a number of factors make wild food plants promising and convenient candidates as new crops. Because they are usually more tolerant than crop plants to adverse climatic and edaphic conditions [9,10] and to pests and diseases [11], they can lead to agricultural systems which are more sustainable and resilient to climate change [12]. From a chemical point of view, a high content of vitamins and minerals is typical of wild greens [13–15]. In some cases, the amount even exceeds that found in cultivated vegetables or other food, which are considered typical sources of that specific nutrient. For example, *Bunias erucago* L. has an iron content much higher than spinach and meat [16]. Moreover, a wide variety of phytochemicals with antioxidant effects have been reported in many of these species [17], and some contain molecules showing antimicrobial potential [18] or other biological-pharmacological activities [19].

In the particular case of baby leaves, the typical early growth stage would make the introduction of wild species into cultivation, in theory, relatively simpler than for crops that have to reach successive and more complex developmental stages, such as the induction of flowering, the fruit setting, the development of underground organs, the formation of the head, etc., thus requiring complicated domestication processes. The safety issue is also to be considered. The cultivation of wild plants would reduce some health risks that the consumption of specimens collected in the wild can entail. These risks are due to the possible accumulation of pollutants such as nitrates and heavy metals, often present in high quantities in the soils in which they live [20,21]. Finally, it is important to remember that wild plants used for food represent a bio-cultural heritage that agriculture could help to safeguard from the risks of progressive depletion linked to the ongoing disappearance of the rural society [5].

The characterization of potential candidates for agricultural exploitation may be the first step for the introduction of wild species into cultivation. The aim of this research was to characterize four common wild edible leafy species (*T. campylodes*, *Rumex acetosa* L., *C. Intybus*, and *Lactuca serriola* L.) from an agronomic, chemical, nutritional, safety, and sensory point of view in order to preliminary evaluate their suitability as baby leaf crops. The first three were chosen for their popularity, and because they had already been grown as minor species [22], *L. serriola* was selected as the ancestor of lettuce, the most important baby leaf crop. In addition, a criterion of choice was the availability of the seeds.

2. Materials and Methods

2.1. Plant Material, Growing Conditions, and Data Collection

Plants of dandelion (*Taraxacum campylodes* G.E.Haglund), sorrel (*Rumex acetosa* L.), wild chicory (*Cichorium intybus* L.), and wild lettuce (*Lactuca serriola* L.) were hydroponically grown in a floating system up to the baby leaf stage (Figure 1).

Figure 1. The four edible wild species investigated in the study: (**a**) dandelion, (**b**) sorrel, (**c**) wild chicory, and (**d**) wild lettuce.

Seeds used as starting material were purchased from "B & T World seeds" (Aigues-vives, France), Provencemonamour (Paris, France), and Fratelli Ingegnoli (Milan, Italy), respectively, for sorrel, wild chicory, and wild lettuce, while seeds of dandelion were collected in the wild in an uncultivated peri-urban area of Lucca (Central Italy) in spring 2021.

Seeds were sown in polystyrene cell trays (L 227.5 mm × W 130.8 mm; 28 cells, Ø 27 mm) filled with vermiculite (Perlite Italiana srl, Corsico, Milan, Italy). Four seeds per cell were sown for each species (3840 seed m^2). After sowing, trays were placed in the dark in a germination chamber at 20 °C for 48 h. Afterward, trays were put in polyethylene terephthalate tanks (L 260.0 mm × W 180.0 mm × H 80 mm) containing 1.8 L of standard Hoagland's nutrient solution [23] prepared with distilled water (macronutrients in mM and micronutrients in μM: 15 N, 1 P, 6 K, 5 Ca, 2 Mg, 50 Fe, 46.2 B, 9.2 Mn, 0.78 Zn, 0.32 Cu, 0.12 Mo; pH 5.52, electric conductivity (EC) 1.1 mS/cm) and moved into a walk-in growth chamber. Here, plants were grown at 24 ± 2 °C (day) and 17 ± 2 °C (night) with a photoperiod of 16 h under fluorescent lighting units OSRAM L36 W/77 (Osram, Munich, Germany). Once a week, tanks were refilled with fresh nutrient solution up to the initial volume.

Plants were harvested 7 weeks after sowing. One part of the harvested material was used to carry out the sensory profile characterization, while another part was used to determine the following parameters: number of plants; fresh weight (FW); dry weight (DW) after oven-drying at 50 °C until a constant weight; leaf area (LA), measured by an area meter LI-3100 (LI-COR, Lincoln, NE, USA); SPAD index, measured immediately before harvest on three fully expanded leaves randomly selected from each tank by means of SPAD-502 chlorophyll meter (Konica-Minolta, Tokyo, Japan); leaf colour, measured immediately after harvest by means of an NR-3000 Portable Colorimeter (Nippon Denshoku Kogyo C., LTD., Tokyo, Japan) according to the CIE system and expressed as L* (lightness or darkness), a* (redness or greenness) and b* (yellowness or blueness) values; and contents of minerals and metals (Al, Ba, Ca, Cd, Cr, Cu, Fe, K, Mg, Mn, Mo, Na, Ni, P, Pb, Se, Sr and Zn), chlorophylls, carotenoids, and nitrate.

2.2. Chemical Analysis

2.2.1. Elemental Composition

Elemental composition was quantified by using 0.5 mg of milled dry sample digested with 10 mL of HNO_3 (67% v/v) in Teflon reaction vessel and then mineralized in a microwave oven (Mars 5, CEM Corp., Matthews, NC, USA) using the program 1600 W, 100% power, at 200 °C for 20 min. At the end of mineralization, the final volume of 25 mL was reached by adding ultra-pure water. The concentrations of Al, Ba, Ca, Cd, Cr, Cu, Fe, K, Mg, Mn, Mo, Na, Ni, P, Pb, Se, Sr, and Zn were determined using an inductively coupled argon plasma optical emission spectrometer (ICP–OES iCAP series 7000 Plus Thermo Scientific, Waltham, MA, USA). A standard method for the 18 different elements was applied, using the Qtegra™ Intelligent Scientific Data Solution™ (ISDS), and the wavelengths selected were 394.4 nm for Al, 493.4 nm for Ba, 315.9 nm for Ca, 228.8 for Cd, 267.7 for Cr, 324.8 nm for Cu, 259.9 nm for Fe, 766.5 nm for K, 285.2 nm for Mg, 257.6 nm for Mn, 204.6 nm for Mo, 589.6 nm for Na, 231.6 nm for Ni, 178.8 nm for P, 220.4 nm for Pb, 196.1 nm for Se, 421.6 nm for Sr, and 206.2 nm for Zn quantification. The calibration was performed with several dilutions of the multi-element standard Astasol®-Mix (ANALYTIKA®, spol. s.r.o., Prague, Czech Republic) in 1% HNO_3 (*v*/*v*) at different concentrations (0.1, 1, 10 and 100 mg/L) with the exception of P, for which the calibration was performed at a P concentration of 10 and 100 mg/L. Blanks and appropriate certified reference materials ERM®-CD281 (B-2440 Geel, Belgium) were included in each batch digested for quality control. Measurements were performed in triplicate. Data were expressed on an FW basis considering the fresh weight/dry weight ratio.

2.2.2. Chlorophyll and Carotenoids

Chlorophylls and carotenoids were extracted from fresh tissues (about 50 mg) using methanol 99.9% as solvent. Samples were kept in a dark room at 4 °C for 24 h, and the assays were carried out immediately after extraction. Chlorophyll *a* (Chl *a*) and chlorophyll *b* (Chl *b*) were determined by the increase in absorbance at 665.2 nm and 652.4 nm, respectively. Total chlorophyll content (Chl *ab*) was determined as the sum of Chl *a* and Chl *b*. Carotenoid content was computed by the increase in absorbance at 470 nm. The pigment concentrations were calculated by Lichtenthaler's formula [24].

2.2.3. Nitrate

Total nitrate content was determined by spectrophotometer using the salicylic acid method [25]. The analysis was performed on samples of 100 mg of dried powdered leaf tissue (80 °C for 48 h) suspended in 10 mL of deionized water on an orbital shaker at room temperature for 2 h. Subsequently, 70 µL of aqueous extract was mixed with 300 µL of 5% salicylic acid in sulphuric acid and with 10 mL of 1.5 M NaOH. The solution was cooled at room temperature for 20 min before reading the absorbance at 410 nm, and the nitrate concentration was calculated through a KNO_3 standard calibration curve. Data were expressed on an FW basis considering the fresh weight/dry weight ratio.

2.3. Contribution to Mineral Requirement and Health Risk Assessment

For evaluating how much the studied species can contribute to human mineral requirements, the amount of mineral elements potentially taken through the consumption of the different baby leaves was calculated as Estimated Dietary Intake (EDI, mg/day) using the following formula:

$$\mathrm{EDI} = \mathrm{C}_{\mathrm{mineral}} \times (\mathrm{CP}/1000) \tag{1}$$

where $C_{mineral}$ is the element concentration (mg/kg FW) in the produce and CP is the consumed portion of baby leaves per day per person, which was assumed to be 50 g [26]. Then, EDI was expressed as percentage (EDI%) of the recommended dietary intake (RDI, mg/day) (for Ca, Cu, Fe, K, Mg, Mo, Na, P, Se, and Zn) or adequate intake (AI, mg/day) (for Cr and Mn) as defined by Italian Society of Human Nutrition (SINU), considering RDI and AI values referred to an adult male [27].

In order to assess the possible health risk due to the intake of heavy metals related to consumption of the baby leaves, the health risk index (HRI) was calculated for the metals detected in leaves of the investigated species according to the following formula:

$$\mathrm{HRI} = \mathrm{EDI}_{\mathrm{Bw}}/\mathrm{RfD} \tag{2}$$

where EDI_{Bw} is the EDI (as defined above) per kg of body weight (BW) and RfD (mg/kg BW/day) is the oral reference dose, which is an estimate of the daily exposure of humans to heavy metals having no hazardous effect during the lifetime according to US-EPA [28]. Since the US-EPA database currently lacks an RfD for Al and Cu, the possible health risk was evaluated on the basis of the tolerable weekly intake (TWI; mg/kg BW/week) of

Al reported by EFSA [29] and Cu RfD according to Taylor et al., 2023 [30]. For BW, an average body weight for an adult was considered and assumed to be 70 kg, as in previous studies [31].

2.4. Sensory Evaluation

The sensory evaluation was conducted by applying the Consensus Profile method as described in ISO 13299:2016 Standard [32] and ISO 5492:2008 Standard [33] at the Mérieux NutriSciences Lab (Prato, Italy).

A group of three trained assessors with broad experience in the sensory evaluation of leafy vegetables were selected to elicit sensory attributes characterizing the products and to assign them the intensity on a 1–5 quantitative scale. Ten attributes were elicited and divided into four sensory modalities: appearance (green colour); odour (herbaceous); flavour (sour, sweet, bitter, and herbaceous aroma); and texture (crunchiness, chewiness, chilliness, and astringency). The selected attributes, their definition, and the intensity scale are shown in Table 1.

Table 1. Attributes, their definitions, and intensity scale used to characterize baby leaves of dandelion, sorrel, wild chicory, and wild lettuce.

Attribute	**Attribute Definition**	**Intensity Scale**
Green colour	Assessment of the green colour tone of the products.	1 = light green; 5 = dark green
Herbaceous odour	Intensity of the odour attributable to the herbaceous and green notes perceived directly by the olfactory system.	1 = absent; 5 = high
Sour	Basic taste typical of organic acid (i.e., citric acid) perceptible inside the oral cavity.	1 = absent; 5 = high
Sweet	Basic taste typical of sugar (i.e., sucrose) perceptible inside the oral cavity.	1 = absent; 5 = high
Bitter	Basic taste typical of caffeine and quinine perceptible inside the oral cavity.	1 = absent; 5 = high
Herbaceous aroma	Intensity of the flavour attributable to herbaceous and green notes perceived indirectly by the olfactory system.	1 = absent; 5 = high
Crunchiness	Property linked to the modality of deformation of the product and to the intensity of the characteristic sound generated during the breaking phase. The product breaks and reproduces the characteristic sound.	1 = low; 5 = high
Chewiness	Characteristic that measures the deformation capacity of the product following its compression and evaluates the ability of the product to return to its original shape without breaking.	1 = low; 5 = high
Chilliness	Burning sensation perceived in the throat or diffusely in the oral cavity.	1 = absent; 5 = high
Astringency	A sensation characterized by contraction of the gums, increase of dryness and roughness on the tongue, and marked decrease in salivation, which commonly occurs by eating unripe fruits.	1 = absent; 5 = high

Immediately before the sensory evaluation session, products were washed and dried by centrifuge. Then, an anonymous sample (identified by a three-digit number) of 100 g for each species was put in a saucer and randomly administered to each assessor. Assessors individually evaluated one sample at a time, recording the intensity of each sensory attribute. At the end of the sensory evaluation session of a single product, the results were collected by the panel leader, which led to a general discussion to reach a mutually agreed consensus for the profile definition. This procedure was repeated until all products were evaluated. Products were evaluated in one replicate. Statistical analysis is not required for this method.

2.5. Experimental Design and Statistical Analysis

Twelve replicates per species (1 replicate = 1 tank) were arranged in a completely randomized block design. Data were expressed as mean ± standard deviation and subjected to one-way analysis of variance (ANOVA). Significant differences between means (n = 3–9) were calculated by the post hoc Tukey's test at $p < 0.05$ using CoStat Software (Version 6.45, Monterey, CA, USA).

3. Results

3.1. Crop Production and Quality

A low emergence was observed in all the species tested in this study. At harvest, seedling emergence percentage was 13.4%, 34.2%, 36.6%, and 43.0% for dandelion, wild lettuce, sorrel, and wild chicory, respectively. Differences in yield were noticed between the species, with sorrel and wild chicory showing higher yields than dandelion and wild lettuce (Figure 2a). On the other hand, dandelion showed higher FW per plant than wild lettuce (Figure 2b), the largest LA (Figure 2c), and a higher DW% than sorrel (Figure 2d).

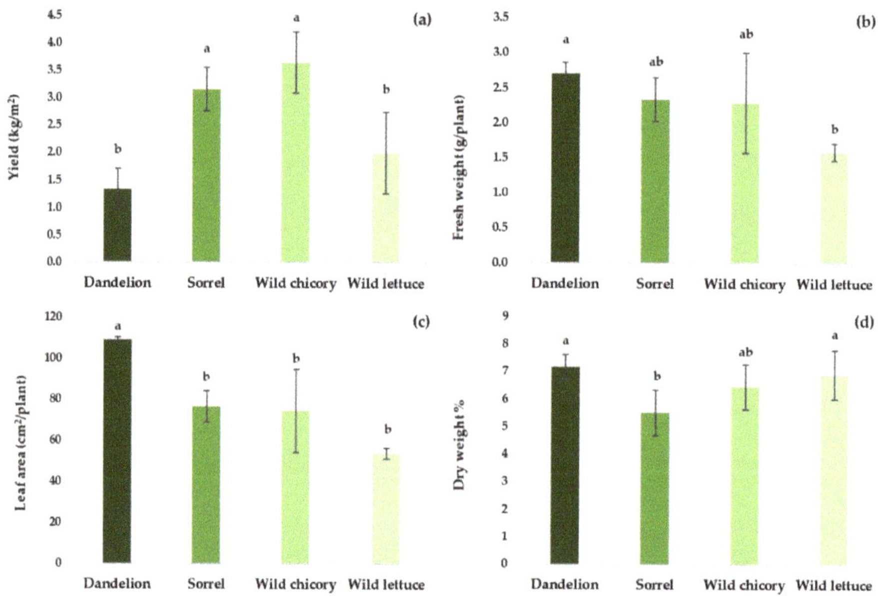

Figure 2. Yield (**a**), fresh weight (**b**) and leaf area per plant (**c**), and DW% (**d**) of baby leaves of the four investigated species (means ± standard deviation). Different letters indicate significant differences according to the post hoc Tukey test ($p < 0.05$).

Wild chicory exhibited higher SPAD values than the other species (Table 2). For what concerns colour parameters, wild lettuce was characterized by the lowest saturation of green (a*), higher saturation of yellow (b*) than sorrel, and lower brightness (L*) in comparison with dandelion and wild chicory (Table 2). The highest contents in Chl *a* and *b* and carotenoids were found in wild chicory, while the lowest was in sorrel (Figure 3a–c). Total chlorophyll content followed the same trend of Chl *a* and Chl *b* (data not shown). Sorrel also showed the lowest nitrate value (926 mg/kg FW), which was 3.2, 4.5, and 5.6 times lower than that of wild lettuce (3005 mg/kg FW), wild chicory (4193 mg/kg FW), and dandelion (5227 mg/kg FW), respectively (Figure 3d).

Table 2. SPAD values and colour parameters of baby leaves of the four investigated species (means ± standard deviation).

	Dandelion	Sorrel	Wild Chicory	Wild Lettuce
SPAD	35.55 ± 2.69 b	33.01 ± 4.17 b	40.10 ± 4.03 a	35.17 ± 3.50 b
Colour parameters				
a*	−7.98 ± 0.56 b	−8.74 ± 0.67 b	−7.74 ± 0.55 b	−3.40 ± 1.31 a
b*	26.73 ± 0.49 ab	23.03 ± 0.69 b	27.26 ± 3.70 ab	29.61 ± 2.69 a
L*	42.23 ± 0.90 a	40.80 ± 2.27 ab	41.74 ± 1.80 a	38.03 ± 1.03 b

Different letters in the same column indicate significant differences ($p < 0.05$) according to the post hoc Tukey test.

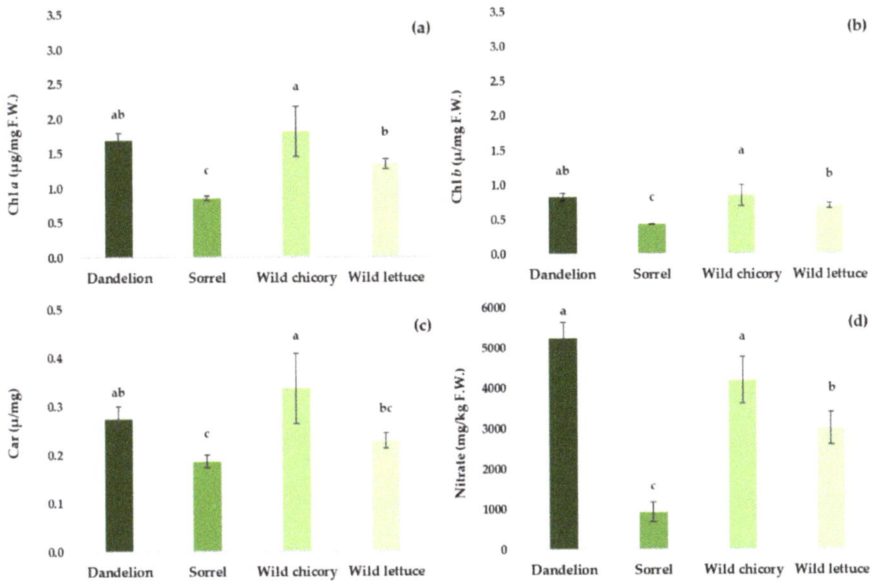

Figure 3. Chlorophylls (Chl *a* and Chl *b*) (**a**,**b**), carotenoids (Car) (**c**), and nitrate content (**d** in baby leaves of the four investigated species (means ± standard deviation). Different letters indicate significant differences according to the post hoc Tukey test ($p < 0.05$).

Lead and Cd were not detected by ICP analysis, and no difference in Al, Ba, Cr, Fe, Mn, Ni, P, Se, and Zn content was found. On the contrary, the investigated species showed significant differences in Ca, Cu, K, Mg, Mo, Na, and Sr concentrations (Table 3). Dandelion was the richest in Cu, Mo, and Sr and contained a higher amount of Ca and K than sorrel. Sorrel showed a greater concentration of Mg than dandelion and wild lettuce. The maximum amount of Na was detected in wild lettuce and the minimum in sorrel.

Table 3. Minerals and metals content of baby leaves of dandelion, sorrel, wild chicory, and wild lettuce (means ± standard deviation).

Minerals and Metals (mg/kg F.W.)	Dandelion	Sorrel	Wild Chicory	Wild Lettuce
Al	9.22 ± 5.14 b	9.45 ± 6.46 a	2.26 ± 0.68 a	3.13 ± 2.03 a
Ba	0.78 ± 0.21 a	1.16 ± 0.34 a	0.72 ± 0.05 a	0.79 ± 0.33 a
Ca	590.81 ± 42.48 a	225.07 ± 18.46 b	467.01 ± 99.29 a	524.46 ± 49.31 a
Cr	0.14 ± 0.09 a	0.21 ± 0.08 a	0.47 ± 0.21 a	0.26 ± 0.16 a
Cu	1.03 ± 0.06 a	0.40 ± 0.04 c	0.44 ± 0.04 bc	0.61 ± 0.11 b
Fe	16.92 ± 6.16 a	9.74 ± 4.03 a	9.15 ± 2.27 a	8.30 ± 1.19 a
K	3330.59 ± 95.22 a	2714.08 ± 86.52 b	3294.62 ± 163.31 a	3554.13 ± 271.03 a
Mg	426.74 ± 46.08 b	565.19 ± 17.60 a	474.31 ± 53.44 ab	422.03 ± 8.97 b
Mn	2.78 ± 0.25 a	5.30 ± 0.14 a	4.93 ± 0.86 a	5.69 ± 2.74 a
Mo	0.15 ± 0.06 a	0.03 ± 0.01 b	0.04 ± 0.01 b	0.02 ± 0.00 b
Na	107.77 ± 13.94 bc	39.52 ± 3.31 c	254.91 ± 11.42 ab	321.90 ± 134.27 a
Ni	0.04 ± 0.02 a	0.10 ± 0.05 a	0.22 ± 0.11 a	0.09 ± 0.10 a
P	409.56 ± 29.12 a	459.39 ± 40.68 a	306.09 ± 23.87 a	480.54 ± 145.93 a
Se	0.18 ± 0.01 a	0.10 ± 0.02 a	0.16 ± 0.05 a	0.12 ± 0.02 a
Sr	2.24 ± 0.18 a	0.38 ± 0.08 c	0.71 ± 0.04 bc	1.140.51 b
Zn	2.56 ± 0.45 a	2.55 ± 0.51 a	1.76 ± 0.57 a	2.61 ± 0.84 a

Different letters in the same row indicate significant differences ($p < 0.05$) according to the post hoc Tukey test.

3.2. Contribution to Mineral Dietary Intake and Health Risk Assessment

The EDI% values resulting from consuming a portion of 50 g a day of baby leaves of dandelion, sorrel, wild chicory, and wild lettuce are listed in Table 4.

Table 4. Estimated dietary intake expressed as a percentage (EDI%) of the recommended dietary intake (RDI) or adequate intake (AI) resulting from the consumption (50 g per day) of baby leaves of dandelion, sorrel, wild chicory, and wild lettuce.

Mineral	**RDI**/*AI* [a] (mg day^{-1})	Dandelion	Sorrel	Wild Chicory	Wild Lettuce
Ca	**1000**	2.95	1.13	2.34	2.62
Cr	*0.035*	19.35	29.49	66.54	37.28
Cu	**0.9**	5.71	2.23	2.42	3.38
Fe	**10**	8.46	4.87	4.57	4.15
K	**3900**	4.27	3.48	4.22	4.56
Mg	**240**	8.89	11.77	9.88	8.79
Mn	*2.7*	5.15	9.81	9.13	10.53
Mo	**0.045**	11.34	2.39	3.04	1.41
Na	**1500**	0.36	0.13	0.85	1.07
P	**700**	2.93	3.28	2.19	3.43
Se	**0.055**	16.70	9.44	14.13	11.32
Zn	**11**	1.07	1.06	0.73	1.09

[a] RDI (bold) and AI (italic), according to SINU (2014).

The highest contribution to RDI/AI was reached for Cr by wild chicory, while the lowest for Na by sorrel. However, differences in the contribution to human mineral requirements between the studied species were not remarkable. Considering the average values, the four investigated species contributed to RDI/AI as follows: 0.6% Na, 1.0% Zn, 2.3% Ca, 3.0% P, 3.4% Cu, 4.1% K, 4.5% Mo, 5.5% Fe, 8.7% Mn, 9.8% Mg, 12.9% Se, and 38.2% Cr.

Table 5 shows the EDI_{bw} and HRI of metals. All the EDI_{bw} values were below the recommended RfD, and HRI was far lower than 1. As regards Al, the weekly consumption of 50 g of product per day would not lead to exceeding the TWI limit of 1 mg/kg body weight/week. The calculated values of weekly consumption were 0.05 mg/kg body weight/week for dandelion and sorrel, 0.01 mg/kg body weight/week for wild chicory, and 0.02 mg/kg body weight/week for wild lettuce.

Table 5. Estimated daily intake per kg of body weight (EDI_{BW}, mg/kg body weight [(1)] day^{-1}) and health risk index (HRI) resulting from the consumption (50 g per day) of baby leaves of dandelion, sorrel, wild chicory, and wild lettuce.

Metal		Dandelion	Sorrel	Wild Chicory	Wild Lettuce
Ba (*Rf*D [(2)] = 0.2)	**EDI_{BW}**	0.000561	0.000827	0.000517	0.000566
	HRI	0.002803	0.004135	0.002586	0.002828
Cr (*Rf*D = 0.003)	**EDI_{BW}**	0.000097	0.000147	0.000333	0.000186
	HRI	0.032254	0.049154	0.110908	0.062130
Cu (*Rf*D = 0.04)	**EDI_{BW}**	0.000734	0.000287	0.000312	0.000435
	HRI	0.018343	0.007165	0.007794	0.010864
Fe (*Rf*D = 0.7)	**EDI_{BW}**	0.012085	0.006957	0.006533	0.005930
	HRI	0.017265	0.009938	0.009334	0.008471
Mn (*Rf*D = 0.14)	**EDI_{BW}**	0.001986	0.003784	0.003523	0.004062
	HRI	0.014186	0.027027	0.025165	0.029018
Mo (*Rf*D = 0.005)	**EDI_{BW}**	0.000105	0.000022	0.000028	0.000013
	HRI	0.021066	0.004437	0.005652	0.002622
Ni (*Rf*D = 0.02)	**EDI_{BW}**	0.000027	0.000073	0.000156	0.000061
	HRI	0.001357	0.003668	0.007776	0.003074

Table 5. *Cont.*

Metal		Dandelion	Sorrel	Wild Chicory	Wild Lettuce
Se (RfD = 0.005)	EDI_{BW}	0.000131	0.000074	0.000111	0.000089
	HRI	0.026247	0.014832	0.022199	0.017788
Sr (RfD = 0.6)	EDI_{BW}	0.001599	0.000268	0.000507	0.000817
	HRI	0.002665	0.000447	0.000844	0.001362
Zn (RfD = 0.3)	EDI_{BW}	0.001599	0.001821	0.001258	0.001862
	HRI	0.006096	0.006071	0.004192	0.006207

(1) Body weight = 70 kg; (2) RfD = oral reference dose (mg kg^{-1} body weight day^{-1}) according to Barnes and Dourson, 1988 [28].

3.3. Sensory Evaluation

The intensity of the 10 attributes elicited by assessors for describing the products is listed in Table 6.

Table 6. Score assigned to each attribute elicited to characterize baby leaves of dandelion, sorrel, wild chicory, and wild lettuce.

Attribute	Dandelion	Sorrel	Wild Chicory	Wild Lettuce
Green colour	4.5	4.5	4.0	4.0
Herbaceous odour	4.0	2.5	4.0	4.5
Sour	1.0	4.5	1.0	1.0
Sweet	1.0	1.5	1.0	3.0
Bitter	4.5	2.0	5.0	3.0
Herbaceous aroma	4.5	2.5	4.5	5.0
Crunchiness	4.0	4.5	4.0	4.0
Chewiness	3.0	2.0	3.0	3.0
Chilliness	2.0	1.0	3.0	2.0
Astringency	3.0	3.0	2.5	2.0

The sensory attributes were perceived at different intensities in the different species, contributing to the definition of peculiar sensory profiles. Dandelion leaves had a very intense green colour and an intense herbaceous odour. Bitter and herbaceous aromas were the most perceived attributes. A medium–high crunchiness and medium chewiness were perceived by the trained judges. Chilliness was mild, while astringency was perceived at medium intensity. The leaves of sorrel had a very intense green colour with an herbaceous odour perceived at medium–low intensity. Sour, perceived at high intensity, was the prevalent flavour, while bitter and sweet were barely perceptible. As noted for the odour, even regarding aroma, the herbaceous note resulted in medium–low intensity. Sorrel showed high crunchiness and low chewiness. Astringency was perceived at medium intensity. Wild chicory showed intense green-coloured leaves with an intense herbaceous odour. Leaves of wild chicory were very crunchy and showed a medium intensity of chewability. In the mouth, the bitterness was perceived at the highest intensity, followed by an herbaceous aroma. Chilliness and astringency were evaluated as moderately intense. Wild lettuce has intense green-coloured leaves and a very intense herbaceous odour. The herbaceous aroma was intense, while sweetness and bitterness were perceived to be balanced at medium intensity. From a textural point of view, the leaves of wild lettuce were very crunchy and chewable. Astringency and chilly sensation were both evaluated as mild. In particular, sorrel was the only species in which a sour sensation was perceived, and chilliness was absent. No sweet taste was detected in wild chicory and dandelion.

4. Discussion

There is very little information in the literature about the growth, yield, and quality at the baby leaf stage of the wild species investigated in this study. The higher yield obtained in sorrel (3.2 kg FW/m^2) and wild chicory (3.7 kg FW/m^2) compared to dandelion

(1.3 kg FW/m^2) was attributable to the greater number of emerged seedlings being the weight of the single plants similar in the three species. The particularly low emergence of dandelion may be due to unfavorable conditions during seed formation or a loss in germination capacity during conservation, considering that the seeds were collected in the wild two years before the cultivation experiment. In a germination test with dandelion seeds pretreated with 2.2% hypochlorite, Lenzi et al. [34] found a 72% germinability. Using these seeds, the authors obtained a yield of about 3.0 kg/m^2. A yield (1.0 kg/m^2) comparable to that found in this work was achieved by Alexopoulos et al. [35]. Wild lettuce, although having an emergence close to that of sorrel and wild chicory, did not show a difference in yield (2.0 kg FW m^2) in comparison with dandelion due to smaller-sized plants. Sorrel, wild chicory, and wild lettuce showed FW per plant and leaf area similar to those observed by Truschi et al. [36] for the same species. Furthermore, our results were consistent with those found by many authors for baby leaf crops grown in soilless culture [37–39]. It is known that high DW% at harvest increases shelf-life in leafy vegetables [40]. In our study, DW% ranged from about 5.0% in sorrel to 7.0% in dandelion. In ten cultivated species at the baby leaf stage, Colonna et al. [41] found values from 5.3% in spinach to 9.4% in rocket.

Colour is an important food quality attribute and plays a significant role in consumers' perception, acceptance, and choice of products [42]. Moreover, many studies report a significant relationship of chromaticity parameters with chlorophyll and total nitrate concentration [43–46]. These authors suggest the use of SPAD meter and colorimeter as non-destructive but accurate analytical methods to estimate chlorophyll and nitrate content. Comparing wild chicory to wild lettuce, our findings were in accordance with these statements. In fact, the results of the chemical analysis revealed higher concentrations of chlorophyll and nitrates in wild chicory than in wild lettuce, consistently with SPAD and the a* and L* values of the two species.

Chlorophylls and carotenoids are an important key factor in crop productivity and quality as they positively influence photosynthetic capacity, visual appearance, and nutraceutical value due to their antioxidant properties [47]. The wild species investigated in this study showed concentrations in chlorophylls and carotenoids close to those observed in leafy vegetables, which are known to be an excellent source of these compounds [34,48–52].

Leafy vegetables may also provide high amounts of minerals in the diet, representing a useful tool to improve human nutritional quality and health status [53]. No information about the elemental composition of wild lettuce baby leaf is available in the literature, but if compared to cultivated lettuce (*L. sativa* L.) grown at the baby leaf stage [41], wild lettuce resulted in being richer in P, Ca, and Mg. The mineral concentration we detected in dandelion, sorrel, and wild chicory was different from that observed by other authors for these species, both collected in nature and cultivated [34,54,55]. Compared to us, these authors found higher concentrations in Ca, Cu, Fe, K, Mn, Na, and Zn but lower in Mg. However, it is worth considering that the mineral content of wild plants may vary significantly depending upon various factors such as genotype, pedoclimatic conditions, season of collection, and developmental stage [5]. The mineral composition is also influenced by the degree of domestication. Ceccanti et al. [54] assessed the composition of some wild leafy species, including sorrel and chicory, observing significant differences between plants gathered in the wild or cultivated. For instance, soilless cultivation of both sorrel and wild chicory resulted in lower and higher content in Ca and Mg, respectively, in comparison with wild collected plants. On the contrary, Disciglio et al. [55] did not observe differences between collected or cultivated wild chicory for Ca and Mg, but Na was higher in cultivated plants.

In accordance with Regulation (EU) No. 1169/2011, foods can be considered significant sources of mineral elements if they contain at least 3000 mg/kg K, 1200 mg/kg Ca, 563 mg/kg Mg, 1050 mg/kg P, 21 mg/kg Fe, 1.5 mg/kg Cu, 15.0 mg/kg Zn, 3.0 mg/kg Mn, 0.06 mg/kg Cr, 0.083 mg/kg Se, and 0.075 mg/kg Mo. Comparing these amounts with those observed in our study, it is possible to state that the investigated species should be a good source of microelements, especially Cr and Se. In fact, EDI% for these elements, considering a daily consumption of a 50 g portion, reached interesting values, ranging from about 19% (dandelion) to 66% (wild chicory) and from 9% (sorrel) to 17% (dandelion), respectively. Chromium and Se, also called oligoelements or trace elements because they are needed in very low quantities, are of fundamental importance to human health. Chromium facilitates the transport of glucose from the blood to the cells [56], while Se enhances the proper function of the immune system and is linked to cardiovascular diseases and cancer prevention [57]. Considering the macroelements, higher contribution to RDI concerned Mg (9.8 EDI% as the average of the four species), whose content in sorrel reached the reference value for significant sources of minerals according to Regulation (EU) No. 1169/2011.

One of the great concerns related to the consumption of leafy vegetables is their capacity to accumulate nitrates and heavy metals in soil. By passing from the soil to the edible organs, these compounds enter the food chain causing potential human health risks. Nitrates have harmful effects because of their ability to form carcinogenic nitrosamines [58]. The maximum nitrate content allowed for the commercialization of leafy vegetables is set within a broad range (2000–7000 mg NO_3/kg) depending on species, season of harvest, and cultivation system by the Commission Regulation (EC) No. 1258/2011. High variability in nitrate content between species was observed in this study. Sorrel showed a nitrate content very far below the lowest recommended limit; wild lettuce and wild chicory fall within the permissible limit set for lettuce, while dandelion exceeded it. In adult plants of dandelion cultivated in a floating system, Alexopoulos et al. [35] and Alexopoulos et al. [59] found a nitrate content ranging from about 600 mg NO_3/kg to 4400 mg NO_3/kg depending on the pH and EC of the nutrient solution. In Lenzi et al. [34], the nitrate concentration of dandelion baby leaf exceeded 7000 mg/kg FW.

Heavy metals are recognized as a serious threat to human health due to the risk associated with their toxicity to the human body and its proper functions [60]. Accumulation of heavy metals in plants, whether wild or cultivated, depends on several factors, such as metal concentrations in the growth medium and water, metal bioavailability, and environmental conditions [61,62]. Different heavy metals were detected in our baby leaves, including some not directly added to the nutrient solution. Considering that the nutrient solution was prepared with distilled water, it can be hypothesised that these elements were present in trace amounts as contaminants of the fertilizers or of the substrate used for the cultivation [63,64]. Dandelion, sorrel, wild chicory, and wild lettuce are reported in the literature as heavy metal indicators or as hyperaccumulator species [65–68], and this ability could explain the presence of these elements in plant tissues. Nevertheless, the HRI calculated for Ba, Cr, Cu, Fe, Mn, Mo, Ni, Se, Sr, and Zn, considering consumption of 50 g per day of baby leaf, were significantly far lower than 1. Moreover, the calculated weekly consumption of Al did not exceed the TWI recommended by EFSA. Lead and Cd, which are considered highly toxic metals even at very low concentrations, were not detected in the baby leaves. Based on these results, we can conclude that the consumption of the four wild investigated species does not pose health risks in relation to heavy metals, at least in the growing conditions we adopted.

The baby leaves of dandelion, sorrel, wild chicory, and wild lettuce showed a distinctive and peculiar sensory profile, well characterized by specific notes. The basic flavour, the herbaceous note, and the spicy sensation were the sensory areas that most contributed to differentiate them. The bitterness was the dominant attribute of flavour in both dandelion and wild chicory, while sorrel's flavour was characterized by sourness. A balance between sweetness and bitterness was detected in wild lettuce. The crunchiness, which is a relevant feature in salads, was perceived in all the species with high intensity.

5. Conclusions

Dandelion, sorrel, wild chicory, and wild lettuce were found to be promising for baby salad production. Their main features can be related to the yield and antioxidant compound content that were comparable to that of leafy vegetables and to the high contribution of Cr, Se, and Mg to dietary intake. From a health point of view, no health risks due to heavy metal accumulation were observed, and sorrel showed a nitrate content very far below the lowest recommended limit for salads. Furthermore, the sensory evaluation highlighted peculiar differences between species. In dandelion and wild chicory, bitterness was perceived as the dominant flavour, while a sweet taste was absent. Sorrel was the only species in which a sour sensation was sensed, and no chilliness was detected. Wild lettuce showed a balance between sweetness and bitterness. Given these distinctive sensory notes, the investigated species could not only be marketed individually but also as a salad mix in order to meet the consumers' demand for new products. Further research on wild species to be used as baby salads should be encouraged because, given the increasing global population and the growing demand for healthy food, finding alternative human food sources is essential.

Author Contributions: Conceptualization, A.B. and A.L.; methodology, A.B., P.B. and A.L.; software, A.B.; validation, A.B. and A.L.; formal analysis, A.B. and A.L.; investigation, A.B., S.T. and A.L.; data curation, A.B., S.T. and A.L.; writing—original draft preparation, A.B. and A.L.; writing—review and editing, A.B., P.B. and A.L.; supervision, A.B., P.B. and A.L. All authors have read and agreed to the published version of the manuscript.

Funding: This research received no external funding.

Data Availability Statement: The data presented in this study are available on request from the corresponding authors.

Acknowledgments: The authors are thankful to Luisa Andrenelli for carrying out the ICP analysis.

Conflicts of Interest: The authors declare no conflict of interest.

References

1. Di Gioia, F.; Renna, M.; Santamaria, P. Sprouts, microgreens and "baby leaf" vegetables. In *Minimally Processed Refrigerated Fruits and Vegetables*; Springer: Boston, MA, USA, 2017; pp. 403–432. [CrossRef]
2. Grand View Research. Packaged Salad Market Size, Share & Trends Analysis Report by Product (Vegetarian, Non-Vegetarian), by Processing (Organic, Conventional), by Type, by Distribution Channel, by Region, and Segment Forecasts, 2021–2028. San Francisco (CA). 2021. Available online: https://www.grandviewresearch.com/industry-analysis/packaged-salad-market (accessed on 5 April 2023).
3. Saini, R.K.; Ko, E.Y.; Keum, Y.S. Minimally processed ready-to-eat baby-leaf vegetables: Production, processing, storage, microbial safety, and nutritional potential. *Food Rev. Int.* **2017**, *33*, 644–663. [CrossRef]
4. Martínez-Sánchez, A.; Luna, M.C.; Selma, M.V.; Tudela, J.A.; Abad, J.; Gil, M.I. Baby-leaf and multi-leaf of green and red lettuces are suitable raw materials for the fresh-cut industry. *Postharvest Biol. Technol.* **2012**, *63*, 1–10. [CrossRef]
5. Baldi, A.; Bruschi, P.; Campeggi, S.; Egea, T.; Rivera, D.; Obón, C.; Lenzi, A. The renaissance of wild food plants: Insights from Tuscany (Italy). *Foods* **2022**, *11*, 300. [CrossRef] [PubMed]
6. Łuczaj, Ł.; Pieroni, A. Nutritional Ethnobotany in Europe: From Emergency Foods to Healthy Folk Cuisines and Contemporary Foraging Trends. In *Mediterranean Wild Edible Plants: Ethnobotany and Food Composition Tables*; Sánchez-Mata, M.D.C., Tardío, J., Eds.; Springer: New York, NY, USA, 2016; pp. 33–56. [CrossRef]
7. Zargar, S.M.; Salgotra, R.K. *Rediscovery of Genetic and Genomic Resources for Future Food Security*; Springer: New York, NY, USA, 2020. [CrossRef]
8. Molina, M.; Pardo-de-Santayana, M.; Tardío, J. Natural Production and Cultivation of Mediterranean Wild Edibles. In *Mediterranean Wild Edible Plants*; Sánchez-Mata, M.D.C., Tardío, J., Eds.; Springer International Publishing: New York, NY, USA, 2016; pp. 81–107. [CrossRef]
9. Razzaq, A.; Wani, S.H.; Saleem, F.; Yu, M.; Zhou, M.; Shabala, S. Rewilding crops for climate resilience: Economic analysis and de novo domestication strategies. *J. Exp. Bot.* **2021**, *72*, 6123–6139. [CrossRef] [PubMed]
10. Hossain, A.; Maitra, S.; Pramanick, B.; Bhutia, K.L.; Ahmad, Z.; Moulick, D.; Syed, M.A.; Shankar, T.; Adeel, M.; Hassan, M.M.; et al. Wild relatives of plants as sources for the development of abiotic stress tolerance in plants. In *Plant Perspectives to Global Climate Changes*; Academic Press: Cambridge, MA, USA, 2022; pp. 471–518. [CrossRef]
11. Hajjar, R.; Hodgkin, T. The use of wild relatives in crop improvement: A survey of developments over the last 20 years. *Euphytica* **2007**, *156*, 1–13. [CrossRef]
12. Shelef, O.; Weisberg, P.J.; Provenza, F.D. The value of native plants and local production in an era of global agriculture. *Front. Plant Sci.* **2017**, *8*, 2069. [CrossRef]
13. Guil-Guerrero, J.L.; Madrid, P.C.; Isasa, M.E.T. Mineral elements determination in wild edible plants. *Ecol. Food Nutr.* **1999**, *38*, 209–222. [CrossRef]

14. García-Herrera, P.; Sánchez-Mata, M.D.C. The contribution of wild plants to dietary intakes of micronutrients (II): Mineral Elements. In *Mediterranean Wild Edible Plants: Ethnobotany and Food Composition Tables*; Sánchez-Mata, M.D.C., Tardío, J., Eds.; Springer: New York, NY, USA, 2016; pp. 141–171. [CrossRef]
15. Sánchez-Mata, M.D.C.; Matallana-González, M.C.; Morales, P. The contribution of wild plants to dietary intakes of micronutrients (I): Vitamins. In *Mediterranean Wild Edible Plants: Ethnobotany and Food Composition Tables*; Sánchez-Mata, M.D.C., Tardío, J., Eds.; Springer: New York, NY, USA, 2016; pp. 111–139. [CrossRef]
16. Ranfa, A.; Maurizi, A.; Romano, B.; Bodesmo, M. The importance of traditional uses and nutraceutical aspects of some edible wild plants in human nutrition: The case of Umbria (central Italy). *Plant Biosyst.* **2014**, *148*, 297–306. [CrossRef]
17. Barros, L.; Morales, P.; Carvalho, A.M.; Ferreira, I.C.F.R. Antioxidant potential of wild plant foods. In *Mediterranean Wild Edible Plants: Ethnobotany and Food Composition Tables*; Sánchez-Mata, M.D.C., Tardío, J., Eds.; Springer: New York, NY, USA, 2016; pp. 209–232. [CrossRef]
18. Di Venere, D.; Gatto, M.A.; Ippolito, A.; Bianco, V.V. Antimicrobial potential of wild edible herbaceous species. In *Mediterranean Wild Edible Plants: Ethnobotany and Food Composition Tables*; Sánchez-Mata, M.D.C., Tardío, J., Eds.; Springer: New York, NY, USA, 2016; pp. 233–252. [CrossRef]
19. Heinrich, M.; Kerrouche, S.; Bharij, K.S. Recent advances in research on wild food plants and their biological pharmacological activity. In *Mediterranean Wild Edible Plants: Ethnobotany and Food Composition Tables*; Sánchez-Mata, M.D.C., Tardío, J., Eds.; Springer: New York, NY, USA, 2016; pp. 253–269. [CrossRef]
20. Brunetti, G.; Soler-Rovira, P.; Farrag, K.; Senesi, N. Tolerance and accumulation of heavy metals by wild plant species grown in contaminated soils in Apulia region, Southern Italy. *Plant Soil* **2009**, *318*, 285–298. [CrossRef]
21. Sánchez-Mata, D.; Morales, R. The Mediterranean landscape and wild edible plants. In *Mediterranean Wild Edible Plants: Ethnobotany and Food Composition Tables*; Sánchez-Mata, M.D.C., Tardío, J., Eds.; Springer: New York, NY, USA, 2016; pp. 15–31. [CrossRef]
22. Tesi, R. *Orticoltura Mediterranea Sostenibile*; Pàtron Editore: Bologna, Italy, 2010.
23. Hoagland, D.R.; Arnon, D.I. The water culture method for growing plants without soil. *Circ. Calif. Agric. Exp. Stn.* **1950**, *347*, 29–31.
24. Lichtenthaler, H.K. Chlorophylls and Carotenoids: Pigments of Photosynthetic Biomembranes. *Method Enzymol.* **1987**, *148*, 350–382. [CrossRef]
25. Cataldo, D.A.; Maroon, M.; Schrader, L.E.; Youngs, V.L. Rapid colorimetric determination of nitrate in plant tissue by nitration of salicylic acid. *Commun. Soil Sci. Plant Anal.* **1975**, *6*, 71–80. [CrossRef]
26. Elgueta, S.; Fuentes, M.; Valenzuela, M.; Zhao, G.; Liu, S.; Lu, H.; Correa, A. Pesticide residues in ready-to-eat leafy vegetables from markets of Santiago, Chile, and consumer's risk. *Food Addit. Contam. Part B Surveill.* **2019**, *12*, 259–267. [CrossRef] [PubMed]
27. SINU. Livelli di Assunzione di Riferimento di Nutrienti ed Energia, IV Revisione. Società Italiana di Nutrizione Umana. Available online: http://www.sinu.it/html/pag/12-MINERALI.asp (accessed on 18 April 2023).
28. Barnes, D.G.; Dourson, M. Reference Dose (RfD): Description and Use in Health risk Assessments. *Regul. Toxicol. Pharmacol.* **1988**, *8*, 471–486. [CrossRef] [PubMed]
29. EFSA. Safety of aluminium from dietary intake—Scientific Opinion of the Panel on Food Additives, Flavourings, Processing Aids and Food Contact Materials (AFC). *EFSA J.* **2008**, *6*, 754.
30. Taylor, A.A.; Tsuji, J.S.; McArdle, M.E.; Adams, W.J.; Goodfellow, W.L., Jr. Recommended reference values for risk assessment of oral exposure to copper. *Risk Anal.* **2023**, *43*, 211–218. [CrossRef]
31. Esposito, M.; De Roma, A.; Cavallo, S.; Miedico, O.; Chiaravalle, E.; Soprano, V.; Baldi, L.; Gallo, P. Trace elements in vegetables and fruits cultivated in Southern Italy. *J. Food Compos. Anal.* **2019**, *84*, 103302. [CrossRef]
32. ISO. *International Standard 13299. Sensory Analysis. Methodology. General Guidance for Establishing a Sensory Profile. International Organization for Standardization*; Ref. No. ISO 13299:2016 (E); ISO: Genéve, Switzerland, 2016.
33. *ISO 5492:2008*; Sensory Analysis—Vocabulary. ISO (International Organization for Standardization): Geneva, Switzerland, 2008.
34. Lenzi, A.; Orlandini, A.; Bulgari, R.; Ferrante, A.; Bruschi, P. Antioxidant and mineral composition of three wild leafy species: A comparison between microgreens and baby greens. *Foods* **2019**, *8*, 487. [CrossRef]
35. Alexopoulos, A.A.; Assimakopoulou, A.; Panagopoulos, P.; Bakea, M.; Vidalis, N.; Karapanos, I.C.; Petropoulos, S.A. Impact of salinity on the growth and chemical composition of two underutilized wild edible greens: *Taraxacum officinale* and *Reichardia picroides*. *Horticulturae* **2021**, *7*, 160. [CrossRef]
36. Truschi, S.; Baldi, A.; Bruschi, P.; Cacciari, I.; Marvasi, M.; Lenzi, A. Foliar roughness and water content impact on Escherichia coli attachment in baby leafy greens. *Biology* **2023**, *12*, 102. [CrossRef]
37. Fallovo, C.; Rouphael, Y.; Rea, E.; Battistelli, A.; Colla, G. Nutrient solution concentration and growing season affect yield and quality of *Lactuca sativa* L. var. *acephala* in floating raft culture. *J. Sci. Food Agric.* **2009**, *89*, 1682–1689. [CrossRef]
38. Manzocco, L.; Foschia, M.; Tomasi, N.; Maifreni, M.; Dalla Costa, L.; Marino, M.; Cortella, G.; Cesco, S. Influence of hydroponic and soil cultivation on quality and shelf life of ready-to-eat lamb's lettuce (*Valerianella locusta* L. Laterr). *J. Sci. Food Agric.* **2011**, *91*, 1373–1380. [CrossRef] [PubMed]

39. Karnoutsos, P.; Karagiovanidis, M.; Bantis, F.; Chatzistathis, T.; Koukounaras, A.; Ntinas, G.K. Controlled root-zone temperature effect on baby leaf vegetables yield and quality in a floating system under mild and extreme weather conditions. *J. Sci. Food Agric.* **2021**, *101*, 3933–3941. [CrossRef] [PubMed]
40. Min, Q.; Marcelis, L.F.; Nicole, C.C.; Woltering, E.J. High light intensity applied shortly before harvest improves lettuce nutritional quality and extends the shelf life. *Front. Plant Sci.* **2021**, *12*, 615355. [CrossRef] [PubMed]
41. Colonna, E.; Rouphael, Y.; Barbieri, G.; De Pascale, S. Nutritional quality of ten leafy vegetables harvested at two light intensities. *Food Chem.* **2016**, *199*, 702–710. [CrossRef]
42. Pathare, P.B.; Opara, U.L.; Al-Said, F.A.J. Colour measurement and analysis in fresh and processed foods: A review. *Food Bioprocess Tech.* **2013**, *6*, 36–60. [CrossRef]
43. Liu, Y.J.; Tong, Y.P.; Zhu, Y.G.; Ding, H.; Smith, F.A. Leaf chlorophyll readings as an indicator for spinach yield and nutritional quality with different nitrogen fertilizer applications. *J. Plant Nutr.* **2006**, *29*, 1207–1217. [CrossRef]
44. Tuncay, O. Relationships between nitrate, chlorophyll and chromaticity values in rocket salad and parsley. *Afr. J. Biotechnol.* **2011**, *10*, 17152–17159. [CrossRef]
45. Mendoza-Tafolla, R.O.; Juarez-Lopez, P.; Ontiveros-Capurata, R.E.; Sandoval-Villa, M.; Iran, A.T.; Alejo-Santiago, G. Estimating nitrogen and chlorophyll status of romaine lettuce using SPAD and at LEAF readings. *Not. Bot. Horti Agrobot. Cluj-Napoca* **2019**, *47*, 751–756. [CrossRef]
46. Amitrano, C.; Rouphael, Y.; De Pascale, S.; De Micco, V. Modulating vapor pressure deficit in the plant micro-environment may enhance the bioactive value of lettuce. *Horticulturae* **2021**, *7*, 32. [CrossRef]
47. Pérez-Gálvez, A.; Viera, I.; Roca, M. Carotenoids and chlorophylls as antioxidants. *Antioxidants* **2020**, *9*, 505. [CrossRef] [PubMed]
48. Žnidarčič, D.; Ban, D.; Šircelj, H. Carotenoid and chlorophyll composition of commonly consumed leafy vegetables in Mediterranean countries. *Food Chem.* **2011**, *129*, 1164–1168. [CrossRef] [PubMed]
49. Ašimović, Z.; Čengić, L.; Hodžić, J.; Murtić, S. Spectrophotometric determination of total chlorophyll content in fresh vegetables. *God. LXI Broj* **2016**, *66*, 104–108.
50. Bulgari, R.; Baldi, A.; Ferrante, A.; Lenzi, A. Yield and quality of basil, Swiss chard, and rocket microgreens grown in a hydroponic system. *N. Z. J. Crop Hortic. Sci.* **2017**, *45*, 119–129. [CrossRef]
51. Ceccanti, C.; Landi, M.; Incrocci, L.; Pardossi, A.; Venturi, F.; Taglieri, I.; Ferroni, G.; Guidi, L. Comparison of three domestications and wild-harvested plants for nutraceutical properties and sensory profiles in five wild edible herbs: Is domestication possible? *Foods* **2020**, *9*, 1065. [CrossRef] [PubMed]
52. Lee, H.W.; Bi, X.; Henry, C.J. Carotenoids, tocopherols and phylloquinone content of 26 green leafy vegetables commonly consumed in Southeast Asia. *Food Chem.* **2022**, *385*, 132729. [CrossRef]
53. Natesh, H.N.; Abbey, L.; Asiedu, S.K. An overview of nutritional and antinutritional factors in green leafy vegetables. *Hortic. Int. J.* **2017**, *1*, 58–65. [CrossRef]
54. Ceccanti, C.; Brizzi, A.; Landi, M.; Incrocci, L.; Pardossi, A.; Guidi, L. Evaluation of major minerals and trace elements in wild and domesticated edible herbs traditionally used in the Mediterranean area. *Biol. Trace Elem. Res.* **2021**, *199*, 3553–3561. [CrossRef]
55. Disciglio, G.; Tarantino, A.; Frabboni, L.; Gagliardi, A.; Giuliani, M.M.; Tarantino, E.; Gatta, G. Qualitative characterisation of cultivated and wild edible plants: Mineral elements, phenols content and antioxidant capacity. *Ital. J. Agron.* **2017**, *12*, 383–394. [CrossRef]
56. Cheng, H.H.; Lai, M.H.; Hou, W.C.; Huang, C.L. Antioxidant effects of chromium supplementation with type 2 diabetes mellitus and euglycemic subjects. *J. Agric. Food Chem.* **2004**, *5*, 1385–1389. [CrossRef]
57. Shadreck, M.; Mugadza, T. Chromium, an essential nutrient and pollutant: A review. *Afr. J. Pure Appl. Chem.* **2013**, *7*, 310–317.
58. Chamandoost, S.; Moradi, M.F.; Hosseini, M.J. A review of nitrate and nitrite toxicity in foods. *J. Hum. Environ. Health Promot.* **2016**, *1*, 80–86. [CrossRef]
59. Alexopoulos, A.A.; Marandos, E.; Assimakopoulou, A.; Vidalis, N.; Petropoulos, S.A.; Karapanos, I.C. Effect of nutrient solution pH on the growth, yield and quality of *Taraxacum officinale* and *Reichardia picroides* in a floating hydroponic system. *Agronomy* **2021**, *11*, 1118. [CrossRef]
60. Jaishankar, M.; Tseten, T.; Anbalagan, N.; Mathew, B.B.; Beeregowda, K.N. Toxicity, mechanism and health effects of some heavy metals. *Interdiscip. Toxicol.* **2014**, *7*, 60. [CrossRef] [PubMed]
61. Filipovic-Trajkovic, R.; Ilic, Z.S.; Sunic, L.; Andjelkovic, S. The potential of different plant species for heavy metals accumulation and distribution. *J. Food Agric. Environ.* **2012**, *10*, 959–964.
62. Gebeyehu, H.R.; Bayissa, L.D. Levels of heavy metals in soil and vegetables and associated health risks in Mojo area, Ethiopia. *PLoS ONE* **2020**, *15*, 0227883. [CrossRef]
63. Jiao, W.; Chen, W.; Chang, A.C.; Page, A.L. Environmental risks of trace elements associated with long-term phosphate fertilizers applications: A review. *Environ. Pollut.* **2012**, *168*, 44–53. [CrossRef]
64. Gunter, M.E.; Singleton, E.; Bandli, B.R.; Lowers, H.A.; Meeker, G.P. Differentiation of commercial vermiculite based on statistical analysis of bulk chemical data: Fingerprinting vermiculite from Libby, Montana U.S.A. *Am. Mineral.* **2005**, *90*, 749–754. [CrossRef]
65. Simon, L.; Martin, H.W.; Adriano, D.C. Chicory (*Cichorium intybus* L.) and dandelion (*Taraxacum officinale* Web.) as phytoindicators of cadmium contamination. *Water Air Soil Pollut.* **1996**, *91*, 351–362. [CrossRef]
66. Porębska, G.; Ostrowska, A. Heavy metal accumulation in wild plants: Implications for phytoremediation. *Pol. J. Environ. Stud.* **1999**, *8*, 433–442.

67. Czarnowska, K.; Milewska, A. The content of heavy metals in an indicator plant (*Taraxacum officinale*) in Warsaw. *Pol. J. Environ. Stud.* **2000**, *9*, 125–128.
68. Gawęda, M. Heavy metal content in common correl plants (*Rumex acetosa* L.) obtained from natural sites in Małopolska Province. *Pol. J. Environ. Stud.* **2009**, *18*, 213–218.

horticulturae

Article

Nutritional Value of Wild and Domesticated *Sanguisorba minor* Scop. Plant

Costanza Ceccanti [1,2,*], Tiane C. Finimundy [2,3] and Lillian Barros [2,3]

1 Department of Agriculture, Food and Environment, University of Pisa, Via del Borghetto, 80-56124 Pisa, Italy
2 Centro de Investigação de Montanha (CIMO), Instituto Politécnico de Bragança, Campus de Santa Apolónia, 5300-253 Bragança, Portugal; tiane@ipb.pt (T.C.F.); lillian@ipb.pt (L.B.)
3 Laboratório Associado para a Sustentabilidade e Tecnologia em Regiões de Montanha (SusTEC), Instituto Politécnico de Bragança, Campus de Santa Apolónia, 5300-253 Bragança, Portugal
* Correspondence: costanza.ceccanti@agr.unipi.it

Abstract: *Sanguisorba minor* Scop. is a wild edible species distributed in the Mediterranean area and present in numerous traditional food recipes. In the present study, the assessment of nutritional value (ash, carbohydrates, fat, proteins, energy, free sugars, organic acids, tocopherols, fatty acid composition, and minerals) of wild and domesticated *S. minor* plants was performed. Results showed an increase in ash, protein, fat, organic acid, and α-tocopherol content after the plant's domestication. Retention of free sugars, especially sucrose, was observed from wild plants to domesticated ones. However, the cultivated plants reported a higher content of polyunsaturated fatty acids than saturated molecules, and both wild collection and domestication maintained a low ω6/ω3 ratio, confirming the role of this species in the prevention of oxidative and inflammatory processes. This aspect is also suggested by the high α-tocopherol content, a vitamin known for its ability to scavenge free-radical species. Nevertheless, a high oxalic acid content was found in domesticated plants. However, the management of fertilization in open field cultivation can be robust in terms of organic acid and mineral (e.g., calcium) content. Indeed, the most representative macrominerals found in domesticated plants were Ca and Mg. The present study suggests a possible introduction of *S. minor* species in the human diet as a functional food or ingredient by virtue of its high nutritional properties and contents. Moreover, the management of fertilization and domestication might be a solution to maintain/enhance the nutritional profile of this wild species.

Keywords: *Sanguisorba minor*; wild harvest; cultivation; nutritional characteristics

Citation: Ceccanti, C.; Finimundy, T.C.; Barros, L. Nutritional Value of Wild and Domesticated *Sanguisorba minor* Scop. Plant. *Horticulturae* **2023**, *9*, 560. https://doi.org/10.3390/horticulturae9050560

Academic Editor: Massimiliano Renna

Received: 22 March 2023
Revised: 4 May 2023
Accepted: 6 May 2023
Published: 9 May 2023

1. Introduction

Sanguisorba minor Scop. is a wild edible species belonging to the Rosaceae family well known for its edibility, its folkloristic and traditional uses, and its nutritional and nutraceutical properties discovered during the last few years [1,2]. *S. minor* is commonly harvested wild in dry and semi-dry grasslands throughout Europe [3]. In Mediterranean traditional food recipes, wild *S. minor* and other species or sub-species of this genus (e.g., *S. minor* spp. *muricata*, *S. officinalis*) are used as boiled vegetables, in traditional soups, and in "misticanza" salad [4]. Nowadays, the rediscovery of ancient food recipes and culinary and medicinal traditions is improving because of the increase in human dietary deficiencies in the undeveloped world [5]. For these reasons, interest in the nutritional and nutraceutical properties of this wild edible herb is increasing, especially regarding its aerial parts and young shoots, thanks to their edibility [6,7]. Viano et al. [8] reported palmitic (29.1%), linoleic (22.6%), and linolenic (21.4%) acids as the main fatty acids and glutamic and aspartic acids as the main amino acids present in wild-collected *S. minor* ssp. *muricata*. Concerning the mineral composition, Pirhofer-Walzl et al. [9] reported a higher concentration of the macrominerals P^+, Mg^{2+}, K^+, and S and the trace elements Zn^{2+} and B^+ in wild-collected *S. minor* than in common grasses. Karkanis et al. [7] observed a high amount of α-tocopherol (vitamin E). This vitamin is well known as an inhibitor of cyclooxygenase 2 (COX-2) involved in inflammatory processes but also as a free-radical scavenger [10–12].

Considering the significant nutritional and nutraceutical properties of *S. minor* species, which are able to diversify the human diet and satisfy the increasing market demands for functional foods as well as enhance nutritional intake, the domestication of *S. minor* plants could be efficient to meet the different requirements of people from different parts of the world [2,7,13]. For this reason, the possibility of domestication of this species is increasingly studied, and the high germination percentage of *S. minor* makes it a promising plant [6]. Therefore, its suitability for different domestications has recently been analyzed in terms of nutritional and bioactive compounds as well as medicinal properties such as antimicrobial or cytotoxic activities [2,7,14,15]. Domestication is usually able to retain chemical compounds present in wild edible species, even though the adopted cultivation technique and the genetic features of the wild edible plants under investigation are the major characteristics affecting the changes in nutritional composition during the domestication [14,16,17].

A deep knowledge concerning the nutritional composition of wild *S. minor* may lead to characterizing the wild nutritional composition of this species and finding an adequate domestication technique able to retain its nutritional composition. Therefore, the aim of this study was to compare and reveal the differences between wild and domesticated *S. minor* plants in terms of nutritional features. In particular, minerals of this species were analyzed for the first time in this experiment in both wild and domesticated *S. minor*, while fatty acids, free sugars, organic acids, and tocopherols were analyzed using proximate analysis for the first time in wild *S. minor* plants.

2. Materials and Methods

2.1. Plant Material and Experimental Design

Seedlings of *S. minor* were collected in the wild (W) in the Tuscany region (43°43′ N, 10°31′ E, Italy) and, at the same time, Tirrenofruit s.r.l., a wholesaler of vegetable food products, provided domesticated *S. minor* plants cultivated on a local farm (F) located in the same area as the wild collection during the spring of 2019. The growing conditions monitored by a weather station near the experimental parcels were: 22 °C average temperature, 50% humidity, and 71.71 lumens cm^{-2} light intensity with 14 h of light. The growth soil was mainly sandy, and 30 kg of manure per ha was used as fertilizer for the soil. *S. minor* young plants were collected by farmers when the shoots were adequate for edibility. Shoots plus leaves were weighed and oven dried at 60 °C until reaching a constant weight, and the dry matter percentage was calculated. Dry samples were used for the analyses described below. All analyses were carried out in triplicate.

2.2. Nutritional Composition Analysis

2.2.1. Proximate Analysis

Samples were analyzed in terms of macronutrients (proteins, fat, carbohydrates, and ash) as reported by the AOAC procedures [18]. The crude protein (N × 6.25) was determined by the Kjeldahl method (978.04) [AOAC, 2005]. The analyte was referred to as "crude" protein because the method determines N, a component of all proteins. In addition, N from sources different from true proteins was also determined. The ash content (AOAC 930.05) was determined by subjecting the sample to incineration at 600 ± 15 °C for 5 h; the crude fat was determined using a Soxhlet apparatus with petroleum ether (AOAC 920.39); and the total carbohydrate content was estimated by difference. The results were expressed in g per 100 g of dry weight (DW). The total energy was calculated using the following equation: energy (kcal) = 4 × (g proteins + g carbohydrates) + 9 × (g fat). Indeed, proteins and carbohydrates provide 4 kcal g^{-1} energy intake, while fats provide 9 kcal g^{-1} energy intake. The results for total energy were expressed in kcal per 100 g DW.

2.2.2. Fatty Acids

The authors previously described the determination of fatty acids using gas-liquid chromatography with flame ionization detection (GC/FID) on a capillary column. Fatty acids were methylated using a transesterification procedure, where 5 mL of methanol:sulfuric acid:toluene in a 2:1:1 (*v*:*v*) ratio was added to the samples, and the mixture was allowed to react for at least 12 h in a bath at 50 °C and 160 rpm. Following this, 3 mL of deionized water were added to obtain phase separation. The fatty acid methyl esters (FAME) were recovered by adding 3 mL of diethyl ether and shaking in a vortex. Then, the upper phase was passed through a microcolumn of anhydrous sodium sulfate to remove water. Resulting samples were recovered in a vial with Teflon, and prior to injection, samples were filtered using a 0.2 μm nylon filter from Millipore [18].

Fatty acid identification was carried out as reported by Barros et al. [19] using gas chromatography coupled with a flame ionization detector (GC-FID/capillary column, DANI model GC 1000, Contone, Switzerland) and a split/splitless injector, using a Macherey–Nagel column (30 m × 0.32 mm I.D. × 0.25 μm d_f). The column temperature in the oven was programmed as follows: starting at 50 °C, it was held for 2 min, then ramped up at a rate of 10 °C per minute to 240 °C and held for 11 min. The carrier gas used was hydrogen, with a flow rate of 4.0 mL/min (0.61 bar), measured at 50 °C. Split injection at a temperature of 250 °C was performed at a ratio of 1:40. For each analysis, 1 μL of sample was injected into the GC. Fatty acid identification was achieved by comparing the relative retention times of the FAME peaks in the samples with the standard (standard mixture 47885-U; fatty acids C4–C24; Sigma, St. Louis, MO, USA). Results were recorded and processed using CSW 1.7 software (DataApex 1.7; Podohradska, Czech Republic) and expressed as relative percentages of each fatty acid.

2.2.3. Free Sugars

Free sugars were determined as previously reported by Barros et al. [19] using the Internal Standard (IS, melezitose, Sigma-Aldrich, St. Louis, MO, USA) method. Dried sample powder (1 g) was extracted in 1 mL of melezitose (5 mg/mL) and 40 mL of 80% (*v*/*v*) aqueous ethanol at 80 °C for 1.5 h, shaking every 15 min. The resulting suspension was centrifuged (Centurion K24OR refrigerated centrifuge, West Sussex, UK) at 15,000× *g* for 10 min. The supernatant was concentrated at 40 °C under reduced pressure and washed three times with 10 mL of diethyl ether, successively. After concentration at 40 °C in the oven, solid residues were dissolved in water to obtain a final volume of 5 mL and filtered through 0.2 μm Whatman nylon filters. After the filtration, the sugar molecules were analyzed by high-performance liquid chromatography coupled to a refraction index detector (HPLC-RI, Knauer, Smartline system 1000). The chromatographic separation was achieved with an Eurospher 100-5 NH2 column (4.6 mm × 250 mm, 5 mm, Knauer) operating at 35 °C (7971R Grace oven). The used mobile phase was acetonitrile/deionized water, 7:3 (*v*/*v*), at a flow rate of 1 mL/min, and the injection volume was 20 μL. Results were recorded and processed using Clarity 2.4 software (DataApex, Prague, Czech Republic). Quantification was performed using internal standards (d-(−)-Fructose, d-(+)-Glucose, d-(+)-Sucrose, and d-(+)-Trehalose; Sigma, St. Louis, MO, USA), and free sugar concentrations were further expressed in g/100 g of DW calculated by internal normalization of the chromatographic peak area.

2.2.4. Organic Acids

Organic acids were determined as previously reported by Barros et al. [19]. Dried sample powder (about 1 g) was extracted with 25 mL of metaphosphoric acid, stirring the solution at 150 rpm at 25 °C for 20 min. Extracted samples were filtered through Whatman No. 4 paper and, successively, through 0.2 μm nylon filters. The Shimadzu 20A series UFLC (Shimadzu Corporation, Kyoto, Japan) was used for the analysis. Separation was achieved on a Phenomenex SphereClone reverse-phase C18 column (5 μm, 250 mm × 4.6 mm I.D.) thermostatically controlled at 35 °C. Elution was carried out using sulfuric acid (3.6 mM) at a flow rate of 0.8 mL/min. Detection was performed using a PDA at the preferred wavelengths of 215 nm and 245 nm (for ascorbic acid). Detected organic acids were quantified by comparison of the area of their peaks recorded at 215 and 245 nm with calibration curves obtained from commercial standards of each compound: ascorbic acid ($y = 8 \times 10^7x + 55{,}079$; $R^2 = 1$); citric acid ($y = 1 \times 10^6x + 4170.6$; $R^2 = 1$); fumaric acid ($y = 172{,}760x + 52{,}193$; $R^2 = 0.999$); malic acid ($y = 952{,}269x + 17{,}803$; $R^2 = 1$); oxalic acid ($y = 1 \times 10^7x + 96{,}178$; $R^2 = 0.999$); and quinic acid ($y = 601{,}768x + 8853.2$; $R^2 = 1$); all were purchased from Sigma (St. Louis, MO, USA). Results were processed using LabSolutions Multi LC-PDA software (Shimadzu Corporation, Kyoto, Japan), and they were expressed in g/100 g DW.

2.2.5. Tocopherols

Tocopherol determination was carried out as previously reported by Barros et al. [19]. The extraction was performed by adding 100 μL of butyl hydroxy toluene solution in hexane (BHT; 10 mg/mL) and 400 μL of tocol solution (SI; Matreya (State College, PA, USA)) in hexane as an internal standard (50 μg mL) to dried samples (500 mg). The extracted samples were homogenized, first with methanol (4 mL) by vortex mixing (1 min), then with hexane (4 mL) by vortex mixing (1 min), and finally with saturated NaCl aqueous solution (2 mL) by vortex mixing again. The resulting samples were centrifuged at 4000× *g* for 5 min at 10 °C, and the supernatant was transferred to a vial. Samples were re-extracted twice with hexane and then dried under nitrogen steam, dehydrated with anhydrous sodium sulfate, and filtered using 0.2 μm nylon filters from Whatman. Tocopherols were determined using an HPLC system (Knauer, Smartline System 1000, Berlin, Germany) coupled to a fluorescence detector (FP-2020; Jasco, Easton, PA, USA) programmed for excitation at 290 nm and emission at 330 nm. The chromatographic separation was achieved with a Polyamide II normal-phase column (250 mm × 4.6 mm; YMC Waters) operating at 30 °C. The mobile phase used was a mixture of *n*-hexane and ethyl acetate (70:30, *v*/*v*) at a flow rate of 1 mL/min. The identification of α-tocopherol was carried out by chromatographic comparison with an authentic standard and quantification by calibration curve obtained from commercial standards (α-, β-, γ-, and δ-isoforms; St. Louis, MO, USA) using the internal standard methodology. Results were recorded and processed using Clarity 2.4 software (DataApex, Prague, Czech Republic). Results were expressed in mg per 100 g DW.

2.2.6. Minerals

Dried sample powders (0.2 g) were mineralized at 220 °C for 90 min using a solution of HNO_3:$HClO_4$ (2.5:1 *v*/*v*) and an atomic absorption spectrometer (Varian AA 24FS, Australia) was used for the determination of Na^+, K^+, Ca^{2+}, Mg^{2+}, Cu^{2+}, Mn^{2+}, Fe^{2+}, and Zn^{2+}. Results were expressed as mg per g fresh weight (FW) for Na, K, Ca, and Mg and μg per g FW for Cu, Mn, Fe, and Zn.

2.3. Statistical Analysis

Differences between wild (W) and domesticated (F) *S. minor* plants were calculated by the Student's *t*-test ($p < 0.05$) with GraphPad Software (GraphPad, La Jolla, San Diego, CA, USA). The data are expressed as the mean ± standard deviation (±SD) of three replicates.

3. Results and Discussion

3.1. Nutritional Value

Table 1 reports results of nutritional composition in terms of moisture, ash, proteins, fat, carbohydrates, and total energy of W and F *S. minor* plants. The highest content of ash, proteins, and fat was recorded in cultivated plants, whereas the same plants showed the lowest moisture and carbohydrate content than W plants.

Table 1. Proximate composition of wild (W) and domesticated (F) *Sanguisorba minor* plants expressed in dry weight (mean ± SD). Data were analyzed following the Student's *t*-test ($p < 0.05$) between W and F plants.

	Proximate Composition		
	W	**F**	***t*-Student *p*-Value**
Moisture (g/100 g)	75.02 ± 0.04	71.20 ± 1.00	0.0054
Ash (g/100 g)	7.80 ± 0.60	10.10 ± 0.40	0.0491
Proteins (g/100 g)	18.80 ± 0.40	23.10 ± 0.30	0.0063
Fat (g/100 g)	2.72 ± 0.04	4.00 ± 0.30	0.0213
Carbohydrates (g/100 g)	70.70 ± 0.20	62.79 ± 0.40	0.0018
Energy (kcal/100 g)	382.00 ± 3.00	379.20 ± 0.50	0.2507

No significant differences were found in terms of caloric energy provided between W and F *S. minor* plants. However, for future steps in the experimentation of the nutritional value of *S. minor* species, the content of soluble and insoluble dietary fibers, consisting of non-starch polysaccharides and other plant components such as cellulose, resistant starch, resistant dextrins, inulin, lignins, pectins, beta-glucans, and oligosaccharides, could be useful to more correctly calculate the caloric value of this species because the energy conversion factor for soluble sugars is 3.75–4.0 kcal/g, whereas that of commonly eaten foods containing a mixture of fermentable and nonfermentable fibers is 0–1.9 kcal/g, according to the Food and Agriculture Organization (FAO). Moreover, other authors analyzed the compositional analysis of Sanguisorba spp. since this plant is considered promising in terms of nutritional and nutraceutical value. Interestingly, Viano et al. [8], analyzing *S. minor* spp. muricata collected as wild in a French park near Marseille, reported a lower content of fat (1.51%) when compared with W or F *S. minor* plants analyzed in the present experiment. However, Tsugkiev et al. [20], analyzing S. officinalis collected as wild at different sea levels, observed lower ash and protein content and higher fat content when compared with W and F plants of the present experiment, even though these authors found many differences between plants collected from different sea levels (e.g., ash content between 5.8 and 7.6 g/100 g, dependent on sea level increase). Thus, different sea levels as well as all other growth environmental parameters might be the reason for different plant chemical compositions. Differences might also be attributed to pedo-climatic growth conditions and genetic features. Moreover, Karkanis et al. [7], analyzing *S. minor* plants whose seeds had been found on wild plants, observed a similar fat and carbohydrate content to the wild or domesticated plants of the present experiment when cultivated with only peat. In this case, the cultivation with only peat when compared with the cultivation with peat and perlite (independently of the percentage of added perlite) resulted more similar to the cultivation in open field or to wild collection, likely due to the presence of perlite, which can lead to the increase of macro- and micropores of plant growth substrate, enhancing the plant nutrient uptake present in the manure used as fertilizer of growth soil, and consequently modify the chemical composition of plant aerial parts [21].

3.2. Sugar Content

The most abundant free sugar identified in both W and F *S. minor* plants was sucrose (Table 2). Wild plants were not significantly affected by domestication in terms of free sugars. Sucrose and raffinose contents were not significantly different in W and F *S. minor* plants, while fructose and glucose in W plants were slightly higher than those in F plants.

Table 2. Sugar content of wild (W) and domesticated (F) *Sanguisorba minor* plants expressed in dry weight (mean ± SD). Data were analyzed following the Student's *t*-test ($p < 0.05$) between W and F plants.

	Sugar Content (g/100 g)		
	W	**F**	***t*-Student *p*-Value**
Fructose	0.50 ± <0.01	0.30 ± 0.10	0.0492
Glucose	1.00 ± 0.10	0.30 ± 0.10	0.0182
Sucrose	3.61 ± 0.04	4.10 ± 0.20	0.0825
Raffinose	0.31 ± 0.01	0.28 ± 0.02	0.3491
Total sugars	5.42 ± 0.03	4.98 ± 0.20	0.1068

Differently, Karkanis et al. [7] showed that fructose and glucose were the major identified sugars in domesticated *S. minor* plants. In general, the sugar content found in the present experiment was lower than that found by Karkanis et al. [7]. Conversely, Viano et al. [8] reported a lower content of total free sugars in *S. minor* spp. *muricata* collected as wild than that of the samples under investigation, independent of cultivation or wild collection. These results confirm that free sugar content is obviously affected by plant growth conditions in the environment. However, the *S. minor* plants of the present experiment might also be considered a rich source of carbohydrates when cultivated, making them an important parameter for the introduction of a new functional food or ingredient in the human diet, even though a crude fiber determination is necessary to confirm this statement.

3.3. Organic Acid and Tocopherol Content

Organic acid content was reported in Table 3. Oxalic acid was found to be the most representative organic acid in the F *S. minor* plants, followed by citric acid. Organic acid content was significantly affected by domestication. The domesticated plants showed a significantly higher content of all organic acids than W *S. minor* plants, except for malic acid.

Table 3. Organic acid and tocopherol content of wild (W) and domesticated (F) *Sanguisorba minor* plants expressed in dry weight (mean ± SD). Data were analyzed following the Student's *t*-test ($p < 0.05$) between W and F plants.

	Organic Acid Content (g/100 g)		
	W	**F**	***t*-Student *p*-Value**
Oxalic acid	2.64 ± 0.05	11.50 ± 0.70	0.0028
Quinic acid	0.16 ± 0.01	1.70 ± 0.10	0.0007
Malic acid	1.14 ± 0.02	0.79 ± 0.01	0.0027
Citric acid	1.76 ± <0.01	5.94 ± 0.01	<0.0001
Fumaric acid	0.11 ± <0.01	0.50 ± 0.03	0.00256
Total organic acids	5.82 ± 0.06	20.43 ± 0.70	0.0013
	Tocopherols (mg/100 g)		
	W	**F**	***t*-Student *p*-value**
α-Tocopherol (vitamin E)	0.50 ± 0.10	11.00 ± 1.00	0.0001

Other authors confirmed that domestication increased the organic acid content, especially that of oxalic and citric acids [7]. Unfortunately, the higher content of oxalic acid in cultivated plants than that of wild plants is a negative aspect since the high intake of this organic acid in the form of oxalate in the human diet can bind with calcium in the body, forming calcium oxalate crystals and leading to kidney stones in human beings [22]. Moreover, high levels of oxalates in the body can interfere with the absorption of certain minerals, such as calcium and iron, leading to mineral deficiencies [23]. However, the management of the domestication and the fertilization (by manure in the present experiment) might be useful to decrease the content of oxalates in *S. minor* edible parts. Boiling, steaming, or other cooking technologies could also be very useful to reduce oxalic acid content since it is well known that this organic acid is very sensitive to high temperatures [24].

Regarding tocopherol content (Table 3), α-tocopherol was the only isoform observed in both F and W *S. minor* samples. Results reported a higher significant content in F *S. minor* plants when compared with W plants, suggesting the possible role of domestication in enhancing α-tocopherol content. The presence of the only α-tocopherol isoform in *S. minor* species was confirmed by Karkanis et al. [7]. Further studies are necessary to understand the biological mechanism that affects the α-tocopherol content during domestication. A speculation could be performed in terms of oxidative stress induced by domestication and the role of α-tocopherol as a scavenger in the light reactions of the photosynthetic process [25]. The oxidative stress could not be present in the wild collected plants due to the adaptation of plants to wild conditions since the collected plants could be a regrowth of the plant. However, the increase in α-tocopherol content during the cultivation of *S. minor* plants enhances the nutraceutical value of this species as a food or ingredient. Indeed, it has been well known for several years that in the diet, α-tocopherol is a vitamin with excellent biological activity as a free-radical scavenger and that it can serve as a therapeutic drug against free-radical-involved diseases as well as an inhibitor of COX-2 involved in the inflammatory process [10–12].

3.4. Fatty Acids

A few significant differences between W and F *S. minor* plants were found in fatty acid content (Table 4). Specifically, lauric (C12:0), pentadecanoic (C15:0), palmitic (C16:0), stearic (C18:0), and lignoceric (C24:0) acids resulted significantly higher in W *S. minor* plants when compared with domesticated ones, while oleic (C18:1n9) and linoleic (C18:2n6) acids reported higher content in domesticated plants than those collected as wild. However, the main fatty acid identified in the aerial part of *S. minor* was α-linolenic acid (36.28–37.50%), followed by palmitic (25.06–22.03%) and linoleic (11.33–13.40%) acids. Similarly, Karkanis et al. [7] observed α-linolenic acid as the most abundant fatty acid (49.4%) in the aerial part of *S. minor* plants, followed by palmitic (14.6–15.6%) and linoleic (12.9–13.1%) acids. In addition, Viano et al. [8] reported α-linolenic, palmitic, and linoleic acids as the most abundant fatty acids in the aerial part of *S. minor* spp. *muricata* plants collected as wild, confirming our findings despite the different species utilized in both experiments. The differences in terms of fatty acid content can be attributed to many factors, such as different environmental growth conditions, cultivation techniques, or harvest periods [14]. However, the similarities in terms of fatty acid varieties with other experiments suggest that the fatty acid composition is also determined by the plant genus.

Saturated fatty acids (SFA) were mainly found in W *S. minor* plants, while polyunsaturated fatty acids (PUFA) were the predominant fatty acid class present in F *S. minor* plants. Clearly, this result affected the PUFA/SFA ratio. Indeed, the wild plants reported a higher PUFA/SFA ratio when compared with domesticated plants. Due to these differences in fatty acid composition, F plants had a higher nutritional value than W plants, as the PUFA/SFA ratio was higher than 1.0, indicating a predominance of PUFA. In the latter case, both analyzed samples had ω6/ω3 ratios less than 0.45, highlighting a predominance of the ω3 family over the ω6. Guil et al. [26] and Simopoulos [27] reported the importance of both of these ratios since they are associated with the beneficial effects of *S. minor* on the cardiovascular system. Indeed, a very high ω6/ω3 ratio promotes the pathogenesis of many diseases, including cardiovascular diseases, cancer, and inflammatory and autoimmune diseases, while increased levels of ω3 PUFA exert suppressive effects [27].

Table 4. Fatty acids of wild (W) and domesticated (F) *Sanguisorba minor* plants expressed in dry weight (mean ± SD). Data were analyzed following the Student's *t*-test ($p < 0.05$) between W and F plants.

Fatty Acids (%)			
	W	**F**	***t*-Student *p*-Value**
C8:0	0.25 ± 0.02	0.16 ± 0.03	0.0691
C10:0	0.46 ± 0.04	0.37 ± 0.02	0.1027
C12:0	1.24 ± 0.04	0.74 ± 0.04	0.0062
C13:0	nd	1.10 ± 0.04	-
C14:0	3.00 ± 0.10	nd	-
C15:0	0.49 ± 0.02	0.35 ± 0.03	0.0212
C16:0	25.10 ± 0.30	22.23 ± 0.05	0.0069
C16:1	2.60 ± 0.20	2.00 ± 0.40	0.2196
C17:0	0.87 ± 0.03	0.60 ± 0.10	0.1447
C18:0	6.30 ± 0.10	5.00 ± 0.10	0.0048
C18:1n9	5.40 ± 0.10	11.40 ± 0.10	0.0003
C18:2n6	11.33 ± 0.03	13.40 ± 0.10	0.0009
C18:3n3	36.30 ± 0.30	37.50 ± 0.40	0.0739
C22:0	1.90 ± 0.40	1.03 ± 0.04	0.087
C24:0	1.79 ± 0.03	0.96 ± 0.01	0.0004
SFA	41.40 ± 0.10	34.50 ± 0.10	0.0003
MUFA	7.95 ± 0.10	13.40 ± 0.30	0.0017
PUFA	47.61 ± 0.30	50.90 ± 0.50	0.0136
PUFA/SFA	1.15 ± <0.01	1.47 ± 0.02	<0.0001
SFA/MUFA	5.21 ± 0.003	2.58 ± 0.002	<0.0001
ω6/ω3	0.31 ± <0.01	0.36 ±< 0.01	0.0037

nd: not detected. Caprylic acid (C8:0); Capric acid (C10:0); Lauric acid (C12:0); Tridecyclic acid (C13:0); Myristic acid (C14:0); Pentadecanoic acid (C15:0); Palmitic acid (C16:0); Palmitoleic acid (C16:1); Heptadecanoic acid (C17:0); Stearic acid (C18:0); Oleic acid (C18:1n9); Linoleic acid (C18:2n6); α-Linolenic acid (C18:3n3); Behenic acid (C22:0); Lignoceric acid (C24:0). SFA—Saturated fatty acids; MUFA—Monounsaturated fatty acids; PUFA—Polyunsaturated fatty acids; PUFA/SFA is the ratio between PUFA and SFA values; ω6/ω3 is the ratio between C18:2n6 and C18:3n3.

3.5. Macromineral and Trace Element Content

Macrominerals and trace elements useful in the diet found in W and F *S. minor* plants were reported in Table 5. Many significant differences were observed between the F and W plants. Among macrominerals, Na^+ and K^+ were higher in W plants, whereas Ca^{2+} and Mg^{2+} were higher in F plants. Among trace elements, the highest contents were observed in W plants. The most representative trace element was Fe.

Table 5. Macromineral and trace element composition of *Sanguisorba minor*, wild-collected (W) and provided by a local farm (F). Each value is the mean (±SD) of three replicates. Macrominerals (Na^+, K^+, Ca^{2+}, and Mg^{2+}) were expressed in mg/g fresh weight, while trace elements (Cu^+, Mn^{2+}, Fe^{2+}, and Zn^{2+}) were expressed in μg/g fresh weight. Data were analyzed following the Student's *t*-test ($p < 0.05$) between W and F plants.

	Na	**K**	**Ca**	**Mg**	**Cu**	**Mn**	**Fe**	**Zn**
W	1.48 ± 0.13	8.30 ± 0.45	1.59 ± 0.03	0.51 ± 0.02	7.09 ± 0.33	37.06 ± 1.90	56.33 ± 3.42	16.97 ± 0.17
F	0.34 ± 0.02	5.51 ± 0.16	4.52 ± 0.26	1.44 ± 0.07	4.55 ± 0.38	13.23 ± 1.59	39.21 ± 5.68	16.39 ± 0.31
t-Student *p*-value	<0.0001	0.0005	<0.0001	<0.0001	0.0010	<0.0001	0.0107	0.0469

The results of the present experiment were confirmed by Pirhofer-Walzl et al. [9], who reported the high content of K in *S. minor* plants collected in the wild. In addition, Lenzi et al. [28] reported similar results for Ca^+, Mg^{2+}, Fe^{2+}, Cu^{2+}, Mn^{2+}, and Zn^{2+} in domesticated *S. minor* plants, confirming our findings.

The lower amount of trace elements in F plants when compared with the W ones suggested the inefficiency of domestication to maintain the trace element content of the wild plants and, consequently, to consider domestication and the management of fertilization as methods to contribute as new sources of macrominerals in the diet of undeveloped countries.

4. Conclusions

The affect of domestication demonstrated positive results, with an increase in some organic acids, α-tocopherol, and macrominerals.

The present work provides valuable results about the characterization of a little-known plant called *Sanguisorba minor*, which could represent a food plant in the future. For this reason, the use of this species in the human diet could be an introduction to a pool of nutritional elements to be considered, especially by people at risk, in the prevention of cardiovascular and chronic diseases. Given its high nutritional value and its edibility, further research about the use of this species as a food or ingredient in the human diet might be needed.

Author Contributions: Conceptualization, C.C. and T.C.F.; methodology, C.C. and T.C.F.; formal analysis, C.C. and T.C.F.; investigation, C.C. and T.C.F.; writing—original draft preparation, C.C. and T.C.F.; writing—review and editing, L.B.; supervision, L.B.; project administration, L.B.; funding acquisition, L.B. All authors have read and agreed to the published version of the manuscript.

Funding: The authors are grateful to the Foundation for Science and Technology (FCT, Portugal) for financial support to CIMO (UIDB/00690/2020 and UIDP/00690/2020) and SusTEC (LA/P/0007/2021). National funding by FCT, P.I., through the individual scientific employment program-contract for the contracts of L. Barros (CEECIND/01011/2018).

Data Availability Statement: Not applicable.

Conflicts of Interest: The authors declare no conflict of interest.

References

1. Ceccanti, C.; Landi, M.; Rocchetti, G.; Moreno, M.B.M.; Lucini, L.; Incrocci, L.; Pardossi, A.; Guidi, L. Hydroponically grown *Sanguisorba minor* scop.: Effects of cut and storage on fresh-cut produce. *Antioxidants* **2019**, *8*, 631. [CrossRef] [PubMed]
2. Finimundy, T.C.; Karkanis, A.; Fernandes, Â.; Petropoulos, S.A.; Calhelha, R.; Petrović, J.; Soković, M.; Rosa, E.; Barros, L.; Ferreira, I.C.F.R. Bioactive properties of *Sanguisorba minor* L. cultivated in central Greece under different fertilization regimes. *Food Chem.* **2020**, *327*, 127043. [CrossRef] [PubMed]
3. Tausch, S.; Leipold, M.; Poschlod, P.; Reisch, C. Molecular markers provide evidence for a broad-fronted recolonisation of the widespread calcareous grassland species *Sanguisorba minor* from southern and cryptic northern refugia. *Plant Biol.* **2017**, *19*, 562–570. [CrossRef] [PubMed]
4. Guarrera, P.M.; Savo, V. Wild food plants used in traditional vegetable mixtures in Italy. *J. Ethnopharmacol.* **2016**, *185*, 202–234. [CrossRef] [PubMed]
5. Flyman, M.V.; Afolayan, A.J. The suitability of wild vegetables for alleviating human dietary deficiencies. *S. Afr. J. Bot.* **2006**, *72*, 492–497. [CrossRef]
6. Ceccanti, C.; Landi, M.; Benvenuti, S.; Pardossi, A.; Guidi, L. Mediterranean Wild Edible Plants: Weeds or "New Functional Crops"? *Molecules* **2018**, *23*, 2299. [CrossRef]
7. Karkanis, A.C.; Fernandes, Â.; Vaz, J.; Petropoulos, S.; Georgiou, E.; Ciric, A.; Sokovic, M.; Oludemi, T.; Barros, L.; Ferreira, I.C.F.R. Chemical composition and bioactive properties of *Sanguisorba minor* Scop. under Mediterranean growing conditions. *Food Funct.* **2019**, *10*, 1340–1351. [CrossRef]
8. Viano, J.; Masotti, V.; Gaydou, E.M. Nutritional value of Mediterranean sheep's burnet (*Sanguisorba minor* ssp. *muricata*). *J. Agric. Food Chem.* **1999**, *47*, 4645–4648. [CrossRef]
9. Pirhofer-Walzl, K.; Søegaard, K.; Høgh-Jensen, H.; Eriksen, J.; Sanderson, M.A.; Rasmussen, J.; Rasmussen, J. Forage herbs improve mineral composition of grassland herbage. *Grass Forage Sci.* **2011**, *66*, 415–423. [CrossRef]
10. Mallet, J.F.; Cerrati, C.; Ucciani, E.; Gamisans, J.; Gruber, M. Antioxidant activity of plant leaves in relation to their alpha-tocopherol content. *Food Chem.* **1994**, *49*, 61–65. [CrossRef]
11. Engin, K.N. Alpha-tocopherol: Looking beyond an antioxidant. *Mol. Vis.* **2009**, *15*, 855. [PubMed]
12. O'Leary, K.A.; De Pascual-Tereasa, S.; Needs, P.W.; Bao, Y.P.; O'Brien, N.M.; Williamson, G. Effect of flavonoids and Vitamin E on cyclooxygenase-2 (COX-2) transcription. *Mutat. Res.-Fundam. Mol. Mech. Mutagen.* **2004**, *551*, 245–254. [CrossRef] [PubMed]
13. Welch, R.M.; Graham, R.D. A new paradigm for world agriculture: Meeting human needs. Productive, sustainable, nutritious. *Field Crops Res.* **1999**, *60*, 1–10. [CrossRef]

14. Ceccanti, C.; Landi, M.; Incrocci, L.; Pardossi, A.; Venturi, F.; Taglieri, I.; Ferroni, G.; Guidi, L. Comparison of three domestications and wild-harvested plants for nutraceutical properties and sensory profiles in five wild edible herbs: Is domestication possible? *Foods* **2020**, *9*, 1065. [CrossRef] [PubMed]
15. Ceccanti, C.; Finimundy, T.C.; Heleno, S.A.; Pires, T.C.; Calhelha, R.C.; Guidi, L.; Ferreira, I.C.F.R.; Barros, L. Differences in the phenolic composition and nutraceutical properties of freeze dried and oven-dried wild and domesticated samples of *Sanguisorba minor* Scop. *LWT* **2021**, *145*, 111335. [CrossRef]
16. Ceccanti, C.; Brizzi, A.; Landi, M.; Incrocci, L.; Pardossi, A.; Guidi, L. Evaluation of major minerals and trace elements in wild and domesticated edible herbs traditionally used in the Mediterranean area. *Biol. Trace Elem. Res.* **2021**, *199*, 3553–3561. [CrossRef] [PubMed]
17. Guarrera, P.M.; Savo, V. Perceived health properties of wild and cultivated food plants in local and popular traditions of Italy: A review. *J. Ethnopharmacol.* **2013**, *146*, 659–680. [CrossRef]
18. Horwitz, W.; Latimer, G.W., Jr. Official Methods of Analysis of AOAC International, 18th ed.AOAC International: Rockville, MD, USA, 2005.
19. Barros, L.; Pereira, E.; Calhelha, R.C.; Dueñas, M.; Carvalho, A.M.; Santos-Buelga, C.; Ferreira, I.C.F.R. Bioactivity and chemical characterization in hydrophilic and lipophilic compounds of *Chenopodium ambrosioides* L. *J. Funct. Foods* **2013**, *5*, 1732–1740. [CrossRef]
20. Tsugkiev, B.; Gagieva, L.; Cherniavskih, V.; Dumacheva, E.; Korolkova, S. Dynamics of the Accumulation of the Biologically Active Substances *Sanguisorba officinalis* L. Depending on Vertical Zonal. In Proceedings of the 1st International Symposium Innovations in Life Sciences (ISILS 2019), Belgorod, Russia, 10–11 October 2019; Atlantis Press: Dordrecht, The Netherlands, 2019; pp. 60–64.
21. Agarwal, P.; Saha, S.; Hariprasad, P. Agro-industrial-residues as potting media: Physicochemical and biological characters and their influence on plant growth. *Biomass Convers. Biorefin.* **2021**, 1–24. [CrossRef]
22. Trinchieri, A. Diet and renal stone formation. *Minerva Med.* **2013**, *104*, 41–54.
23. Gemede, H.F.; Haki, G.D.; Beyene, F.; Woldegiorgis, A.Z.; Rakshit, S.K. Proximate, mineral, and antinutrient compositions of indigenous Okra (*Abelmoschus esculentus*) pod accessions: Implications for mineral bioavailability. *Food Sci. Nutr.* **2016**, *4*, 223–233. [CrossRef] [PubMed]
24. Chai, W.; Liebman, M. Effect of different cooking methods on vegetable oxalate content. *J. Agric. Food Chem.* **2005**, *53*, 3027–3030. [CrossRef] [PubMed]
25. Trebst, A.; Depka, B.; Holländer-Czytko, H. A specific role for tocopherol and of chemical singlet oxygen quenchers in the maintenance of photosystem II structure and function in *Chlamydomonas reinhardtii*. *FEBS Lett.* **2002**, *516*, 156–160. [CrossRef]
26. Guil, J.L.; Torija, M.E.; Giménez, J.J.; Rodríguez, I. Identification of fatty acids in edible wild plants by gas chromatography. *J. Chromatogr. A* **1996**, *719*, 229–235. [CrossRef] [PubMed]
27. Simopoulos, A.P. The importance of the omega-6/omega-3 fatty acid ratio in cardiovascular disease and other chronic diseases. *Exp. Biol. Med.* **2008**, *233*, 674–688. [CrossRef] [PubMed]
28. Lenzi, A.; Orlandini, A.; Bulgari, R.; Ferrante, A.; Bruschi, P. Antioxidant and mineral composition of three wild leafy species: A comparison between microgreens and baby greens. *Foods* **2019**, *8*, 487. [CrossRef]

horticulturae MDPI

Article

Soilless Cultivation of *Portulaca oleracea* Using Medicinal and Aromatic Plant Residues for Partial Peat Replacement

Antonios Chrysargyris [1], Stavros Louka [1], Spyridon A. Petropoulos [2] and Nikolaos Tzortzakis [1,*]

[1] Department of Agricultural Sciences, Biotechnology and Food Science, Cyprus University of Technology, Limassol 3603, Cyprus
[2] Department of Agriculture, Crop Production and Rural Environment, University of Thessaly, 38446 Volos, Greece
* Correspondence: nikolaos.tzortzakis@cut.ac.cy

Abstract: The industrial manufacturing of essential oils (EOs) generates a sizable volume of bulk solid waste (SW) that needs to be disposed of. The present study evaluated the potential of using *Origanum dubium* wastes (ODW) and *Sideritis cypria* waste (SCW) obtained after EO distillation for partial peat substitution (0–5–10–20–40% v/v) in *Portulaca oleracea* production. Both ODW and SCW increased pH, electrical conductivity, organic matter, and mineral content, but negatively affected the total porosity and aeration of the growing media. Plant growth was inhibited, especially when high ratios of residues were used, and this was reflected by leaf stomatal conductance and chlorophyll decrease, as well as by the activation of several nonenzymatic (phenols, flavonoids, and antioxidant capacity) and enzymatic (catalase, superoxide dismutase, and peroxidase) mechanisms and the increase in lipid peroxidation and hydrogen peroxide, indicating stress conditions. Despite that both ODW and SCW were rich in minerals, plants could not accumulate them. It can be concluded that both ODW and SCW have the potential to be used in the growing media at low ratios up to 10%, with increased antioxidant content in the final product. Nonetheless, the growing media properties, i.e., total pore space and aeration, still need to be improved to result in sufficient yields.

Keywords: purslane; distillation waste; plant growth; peat; unexploited vegetables; antioxidants; minerals; total phenols

Citation: Chrysargyris, A.; Louka, S.; Petropoulos, S.A.; Tzortzakis, N. Soilless Cultivation of *Portulaca oleracea* Using Medicinal and Aromatic Plant Residues for Partial Peat Replacement. *Horticulturae* **2023**, 9, 474. https://doi.org/10.3390/horticulturae9040474

Academic Editor: Bastien Christ

Received: 12 March 2023
Revised: 1 April 2023
Accepted: 6 April 2023
Published: 10 April 2023

1. Introduction

The world's agricultural industry produces a significant amount of bioresidues (solid, liquid, and gaseous), which are considered the most abundant, profitable, and renewable resource on the planet [1]. The management and recycling of these residues should be conducted in an ecologically and environmentally friendly way; otherwise, they may result in environmental hazards, expanding our concerns about human health constraints and the environment. Any biological material that is not consciously produced during a production process is referred to as residual biomass [2]. The leftover biomass is referred as a byproduct, which could be a potential waste depending on its management [3]. In the context of economic gain, environmental sustainability, and social advantages, the efficient usage and recycling of residual biomass generated by the medicinal and aromatic plant (MAP) sector are of utmost importance.

MAP material is used raw, dried, frozen for essential oil (EOs) production and for plant extracts (e.g., infusions and decoctions). Depending on the MAP species, the herbal, perfumery, and cosmetic industries produce a variety of solid and liquid leftovers, including the byproducts of aromatic plant distillation and the nonused parts of medicinal plants. After the primary processing, certain medicinal plant components that are specifically employed for medicinal purposes are used in the form of raw pharmaceuticals, whereas the other portions of the plant are left unused and end up as waste [2].

The primary methods used to extract essential oils include hydrodistillation, steam distillation, hydro steam distillation, organic solvent extraction, and mechanical extraction. All of these methods are time and cost demanding to operate [4], while the vast amounts of biomass residues generated from aromatic plant distillation go unused and decrease the added value of the crops [5]. It is estimated that MAP residues account for more than 200,000 tons worldwide annually [6]. Therefore, the efficient recycling of this waste material could provide additional revenue to the producers of aromatic plants. To further valorize the MAPs' generated waste, both conventional and innovative ways have been suggested, and cutting-edge technologies are being developed [7]. Composting is a common process to reuse crop residues, and the obtained composts can be used as organic fertilizers and soil-improving amendments [8–10]. Other uses include distilled lavender stalks that have been implemented as bioaggregate for building material [11], *Aloe vera* waste that is being used in the diet of lactating cows [12], and ginseng residues that are used as dietary feed supplements for piglets [13]. On the other hand, the use of crop residues as growing material in soil-less agriculture, in either composted or raw form, has received less attention so far [14–17].

EO yield can range from very low contents to typically up to 5–8% of the extracted biomass, thus resulting in enormous amounts of residues (solid waste and hydrolate—the aqueous byproduct after distillation), which need to be properly managed [2]. The vast majority of these leftovers are burned and disposed of as waste in landfills [18], posing an adverse effect on the environment, increasing energy cost, and dismantling the circular use of resources, since several components remain in the residues and can be used to obtain valuable bioactive components, such as phenolic compounds [7,13,19]. The exploitation of new natural antioxidants has greatly increased over the last few decades, and still there is plenty of space to explore agroindustrial byproducts as a source of antioxidants. A solution with high potential could be the employment of these residues as an alternative source of antioxidants because of their low cost, high availability, and need for environmentally favorable handling [20].

Peat is the principal growing medium utilized in the agriculture sector due to its useful agronomic characteristics, and as a result, an estimated 14–20% of the used peat is normally allocated to the horticulture industry [21,22]. Peat extraction and use should be gradually decreased as it is a nonrenewable resource and a large CO_2 sink [23]. Several peat alternatives have been searched so far, including the use of several agroindustry wastes as growing media in soil-less or pot-grown ornamental and horticultural crops [14,24–29]. During the growing medium preparation, the material, the ratio, the fertility, and the physicochemical properties of the individual components can be appropriately selected [25,26,30,31]. However, there are several farmers who use unprocessed materials in their fields without considering any possible harmful effects on the crops, the ecosystem, and the general public health [32,33]. The use of fresh rice hull in deciduous tree and flowering shrub cultivation [34], fresh rice and kenaf for the production of *Pinus halepensis* seedlings [35], olive mill waste (OMW) and grape mill waste (GMW) in vegetable production [36], and shredded paper waste [24,37] has been reported, which all present specific limitations for their successive application, depending on the material used, the growing/environmental conditions, and the plant species examined.

Peat can be partially replaced on a commercial scale only when the alternative materials meet or substitute some of the peat's main properties. Purslane is a wild herb, popular in Mediterranean and Asian diets, with high nutritive and pharmacological properties, which attribute its name as "the future superfood" [38–41]. Considering the adaptability of the species to adverse environmental conditions, it is also suggested as a potential alternative crop to mitigate the negative impacts of climate change and soil degradation on crop production [42,43]. Therefore, the main objective of this study was to explore the use of byproducts from the MAP sector to partially replace peat in the production of unexploded vegetables such as purslane (*Portulaca oleracea*). The second objective was to determine how the tested byproducts impacted the nutritional value, chemical content, and bioactive components of purslane leaves to identify the best growing medium that could enhance plant growth parameters and boost the final product's quality.

2. Materials and Methods

2.1. Plant Material and Growing Media Preparation

The present study took place at the hydroponic infrastructures (full climate automatic control plastic greenhouse) at the Cyprus University of Technology, Limassol, Cyprus. Commercial-based peat (professional peat, Gebr. Brill Substrate GmbH & Co. KG, Georgsdorf, Germany) was used as the growing substrate in this study. Sufficient minerals were added in peat by employing common fertilizers (Novatec, simple superphosphate, potassium sulfate) at 75 mg N/L, 22 mg P/L, 104 mg K/L of growing medium. The peat and the fertilizers were thoroughly mixed with a professional concrete mixer.

Medicinal and aromatic plant waste included *Origanum dubium* distilled waste-ODW and *Sideritis cypria* distilled waste-SCW, as derived after the steam hydro distillation process for essential oil extraction from the relevant aerial plant material. MAP material was provided by the Department of Agriculture, Sector of Medicinal and Aromatic Plants, Nicosia, Cyprus. Plants were cultivated under conventional farming practices, such as annual soil tillage, pruning of plants, fertilization, and crop protection applications with pesticides as appropriate for the region and following the best practice guides for the respective crops. MAPs were air-dried in the shade before being put through a 60 L semicommercial distillator for steam hydrodistillation. The distillation residues were left to dry (moisture was < 10%), then shredded and stored under dry conditions until further use.

Peat (P) was used as the base ingredient of the growing medium and was proportionally replaced with various ratios of ODW or SCW, resulting in the following nine media mixtures (*v*/*v*): (1) peat 100% (control), (2) P:ODW 95:5 (ODW 5%), (3) P:ODW 90:10 (ODW 10%), (4) P:ODW 80:20 (ODW 20%), and (5) P:ODW 60:40 (ODW 40%) for the oregano residues and (6) P:SCW 95:5 (SCW 5%), (7) P:SCW 90:10 (SCW 10%), (8) P:SCW 80:20 (GSC 20%) and (9) P:SCW 60:40 (SCW 40%), for the sideritis residues. Raw growing media were collected and analyzed for their physicochemical properties prior to seedling transplantation. The particle size of the shredded dried residues was also determined by using an electromagnetic and digital sieve shaker (BA-200-N for 8 × 200 mm sieves 230 V–50 Hz, CISA, Barcelona, Spain), for particle separation, fraction, and size determination, with a two-dimensional movement (a horizontal, circular motion and a vertical, tapping action), which allows particle stratification. The eight sieve sizes ranged from 4.00 mm to <75 μm. The uniformity index (UI = d_{60}/d_{10}) is a measure of the uniformity of particle size in the growing media and is defined as the ratio of the 60% finer size (d_{60}) to the effective size (d_{10}) (which means that 10% of the particles are finer and 90% of the particles are coarser than d_{10}) [44]. The ODW and SCW ratios were determined according to preliminary tests and/or previous experience with plant residues incorporated in growing media [24,36].

Purslane (*Portulaca oleracea* L.) seeds were bought from Hortus Sementi Srl. (Budrio, Italy; 2020 production lot) and placed in 72-cell black plastic trays filled with commercial peat, under nursery conditions. When seedlings reached the third leaf plant developmental stage, they were transferred into 0.3 L plastic pots with one of the nine different growing media. Eight replicate pots with one seedling per pot were utilized for each treatment (growing media). The pots were placed in plastic trays to preserve the drained solution after each watering. Plants were watered through the plastic trays with capillary suction, based on plants' needs. No fertilizers, insecticides, or other agrochemicals were used during the seedling growth. Throughout the cultivation season, average temperatures of 20.8 °C (Tmin of 16.9 °C; Tmax of 32.8 °C) and relative humidity levels of 57.4% were recorded.

2.2. Growing Media Characteristics

The physicochemical characteristics of the raw materials (P, ODW, and SCW) and the tested growing media were determined. Total pore space (TPS), air-filled porosity (AFP), available water holding capacity (AWHC), and bulk density (BD) by volume of the growing media were investigated based on European Standards, EN 13041 [45], as previously described [24]. The pH and the electrical conductivity (EC) of each growing media were measured in each growing medium extracted with water at a ratio of 1:5 v/v. Organic matter and organic C were determined after media ashing at 550 °C in a furnace [31]. For mineral analysis, the ash samples were then acid-digested (2 N HCl) following the protocol of Chrysargyris et al. [24], while macronutrients, such as potassium (K), sodium (Na), magnesium (Mg), and calcium (Ca), were measured using ion chromatography (ICS-3000, Dionex Aquion, Sunnyvale, CA, USA) and the IonPac CS19 (4 × 250 mm, Dionex, Co., Sunnyvale, CA, USA) analytical column. Phosphorus (P) was determined by spectrophotometry (Multiskan GO, Thermo Fischer Scientific, Waltham, MA, USA), and nitrogen (N) was determined by the Kjeldahl method (BUCHI, Digest Automat K-439 and Distillation KjelFlex K-360, Flawil, Switzerland). Data were expressed in g/kg of dry weight.

2.3. Plant Growth, Physiology, and Mineral Analysis

Purslane plants were grown for 25 days, and various growth parameters were measured in six seedlings per treatment. Seedling height and the number of leaves produced per plant were measured. Seedlings were harvested, and the upper fresh biomass was weighed (g) and dried, and then the total dry weight (g) was measured.

Additionally, several physiological parameters were recorded before harvesting. Leaf stomatal conductance was determined with a ΔT-Porometer AP4 (Delta-T Devices, Cambridge, UK). Leaf chlorophyll fluorescence was recorded on two fully expanded, sun-exposed leaves per plant (Opti-Sciences fluorometer OS-30p, Hertfordshire, UK). Leaf chlorophylls (chlorophyll a-Chl a, chlorophyll b-Chl b, total chlorophylls-total Chl) and carotenoid content were also assessed (six replications/treatment), as described previously, and the results were expressed as mg of chlorophyll (or carotenoids) per gram of fresh tissue [46].

Mineral content in plant leaves was determined in four replications/treatment (two pooled plants/replication) [46]. Plant tissue was ashed in a furnace (Carbolite, AAF 1100, GERO, Lilienthal, Germany) at 480 °C for 6 h and acid-digested (2 N HCl). Minerals (N, K, P, Na, Ca, and Mg) were determined as described above, and results were expressed in g/kg of dry weight [24].

2.4. Total Phenolic Compounds, Total Flavonoids, and Antioxidant Activity

Total phenolic compounds, total flavonoids, and total antioxidant activity were determined in the methanolic extracts obtained from plant tissues (four replicates/two pooled plants per replicate) for each treatment. For the total phenolic compounds, the Folin–Ciocalteu reagent (Merck, Darmstadt, Germany) was used, and results were presented as mg of gallic acid equivalents per g of fresh weight (fw) [47]. Total flavonoid content was determined according to the aluminum chloride colorimetric method [48], and results were presented as rutin equivalents (mg rutin/g of fw). For antioxidant activity, two assays were employed, ferric reducing antioxidant power (FRAP) and 2,2-diphenyl-1-picrylhydrazyl (DPPH), which were performed as described previously by Chrysargyris et al. [24]; results were presented as Trolox equivalents per g of fresh weight.

2.5. Lipid Peroxidation, Hydrogen Peroxide Content, and Enzyme Antioxidant Activity

Lipid peroxidation indicated by the malondialdehyde content (MDA) and hydrogen peroxide (H_2O_2) content was determined as previously described in the literature [49,50]. The results were expressed as nmol of MDA per g of fw and as μmol H_2O_2 per g of fw.

The antioxidant enzyme activity for superoxide dismutase (SOD) (EC 1.15.1.1), catalase (CAT) (EC 1.11.1.6), and peroxidase activity (POD) (EC 1.11.1.6) was determined as described previously [36]. Results were presented as enzyme units per mg of protein. The protein content was determined by the Bradford method, and bovine serum albumin (BSA) was used as standard.

2.6. Statistical Analysis

Data were statistically analyzed by the IBM SPSS v22.0 (SPSS Inc., Chicago, IL, USA) program. Prior to analysis of variance (ANOVA), data were checked for normality. Then raw data for each waste material (ODW or SCW) were subjected to one-way ANOVA, and when significant effects were recorded, the comparison of means was performed with Duncan's multiple range test (DMRT) at $p \leq 0.05$.

3. Results and Discussion

Not all crop residues are ideal for preparing suitable growing media for crop production in pots or for the production of seedlings or cuttings in nurseries of horticultural crops. The acidity, alkalinity, salinity, and presence of phytochemical compounds (such as polyphenols) at excessive levels may restrict the use of these residues due to possible phytotoxic effects [51]. Therefore, a detailed analysis of several physical, chemical, and biological variables is needed to determine the impact of growing media composition on plant growth and yield. The tested raw MAP residues (ODW and SCW) were rich in minerals and contributed to the physicochemical characteristics of the final growing media mixtures when combined at various ratios with peat (Tables 1 and 2). More specifically, ODW had a slide acidic pH (averaged values of 5.96), high organic content, and bulk density, and it contained high levels of N (10.51 g/kg), K (13.46 g/kg), P (2.83 g/kg), Na (1.22 g/kg), and Mg (2.67 g/kg), which resulted in increased EC values, compared with peat (Table 1). However, ODW had almost the half level of Ca (averaged values of 7.66 g/kg) compared with peat. Regarding SCW, it recorded an almost neutral pH (averaged values of 6.75), high organic matter, and available water holding capacity, as well as high levels of N (12.66 g/kg), K (14.06 g/kg), P (1.65 g/kg), Na (5.79 g/kg), and Mg (1.70 g/kg), which also resulted in increased EC values compared with peat (Table 2). On the other hand, SCW contained only lower amounts of Ca (averaged values of 11.58 g/kg) and recorded a lower bulk density compared with peat. In both materials, ODW and SCW revealed a uniformity index (UI = d_{60}/d_{10}) of UI = 20 and UI = 14.4, respectively [44], having 90% of particles with sizes lower than 0.4 mm (resulted in a dustier material) and 3.6 mm (resulted in various size materials), respectively (Figure S1). The tested MAP residues could be considered fertile material, as it has been previously reported that distilled MAP waste may contain micronutrients and macronutrients at amounts of 0.35–1.80% N, 0.45–0.60% P, and 2.00–2.25% K [52]. Due to the wide variety of crop and processing industry residues, there are no references on the properties of the tested raw materials; however, there are references on composted materials. The analyzed ODW and SCW properties were within the range of the available data from other studies, apart from the air-filled porosity of ODW, which was 1.57% [53]. The EC of the studied ODW and SCW materials was very low and below the highest recommended EC of 4 mS/cm (the Greek standard) for composted materials, which excludes any initial phytotoxic effects on young plants [54].

Adding MAP residues in different ratios into the growing media for partial peat substitution also affected the physicochemical properties of the obtained mixtures (Tables 1 and 2). The presence of ODW or SCW increased the growing media pH by almost 1.2 pH unit compared with the control (peat) media, resulting in pH values of 7.51 and 7.54 at ODW 40% media and SCW 40% media, respectively. Indeed, the pH increase recorded in mixtures that contained ODW and SCW at ≥20% ratios was greater than the recommended pH levels for growing media (5.3–6.5) [53]. When ODW was combined with peat at the highest ratio (ODW 40%), it had a considerable impact on the level of organic matter content in the mixture, whereas no significant effects were recorded for SCW when combined with peat (Tables 1 and 2). Considering that agricultural soils in the Mediterranean basin often contain less than 3–4% of organic matter, there is a significant amount of organic matter in the waste biomass generated during MAP processing that could be used as soil or growing medium amendments [2].

Table 1. Growing media (peat, *Origanum dubium* waste—ODW) physicochemical properties before plant transplanting.

	Peat 100%	ODW 5%	ODW 10%	ODW 20%	ODW 40%	ODW 100%
pH	6.32 ± 0.17 b	6.34 ± 0.05 b	6.31 ± 0.09 b	6.62 ± 0.01 b	7.51 ± 0.12 a	5.96 ± 0.07 c
EC (mS/cm)	0.84 ± 0.04 c	1.13 ± 0.07 b,c	0.89 ± 0.00 b,c	1.12 ± 0.04 b	1.70 ± 0.18 a	1.92 ± 0.01 a
Organic matter (%)	72.38 ± 1.28 c,d	73.03 ± 0.36 c	73.30 ± 0.73 c	70.16 ± 0.38 d	76.91 ± 1.08 b	92.79 ± 0.25 a
Organic C (%)	41.98 ± 0.74 c,d	42.36 ± 0.21 c	42.51 ± 0.42 c	40.69 ± 0.22 d	44.61 ± 0.63 b	53.82 ± 0.14 a
C/N ratio	50.36 ± 2.12 a	42.91 ± 0.99 b	40.92 ± 3.35 b	26.22 ± 0.67 c	28.33 ± 0.79 c	51.21 ± 0.23 a
N (g/kg)	8.35 ± 0.19 c	9.88 ± 0.20 b	10.51 ± 0.76 b	15.53 ± 0.33 a	15.77 ± 0.59 a	10.51 ± 0.05 b
K (g/kg)	2.03 ± 0.05 d	3.86 ± 0.31 c	3.96 ± 0.41 c	4.69 ± 0.13 c	7.35 ± 0.31 b	13.46 ± 0.11 a
P (g/kg)	1.12 ± 0.04 c	1.61 ± 0.29 b,c	1.72 ± 0.09 b	1.91 ± 0.18 b	2.62 ± 0.18 a	2.83 ± 0.02 a
Ca (g/kg)	15.01 ± 0.56 b	21.52 ± 2.43 a	17.62 ± 0.98 b	20.41 ± 0.21 a	20.51 ± 0.49 a	7.66 ± 0.27 c
Mg (g/kg)	0.79 ± 0.04 e	1.51 ± 0.21 d	1.50 ± 0.09 d	2.22 ± 0.02 c	3.29 ± 0.06 a	2.67 ± 0.10 b
Na (g/kg)	0.97 ± 0.03 c	1.12 ± 0.12 b	1.12 ± 0.08 a,b	1.24 ± 0.05 a,b	1.31 ± 0.03 a	1.22 ± 0.08 a,b
Total porosity %	84.97 ± 0.76 a	72.68 ± 3.31 b	77.19 ± 4.48 a,b	53.13 ± 1.37 c	48.60 ± 3.67 c	69.87 ± 3.89 b
Air-filled porosity (% *v*/*v*)	18.43 ± 1.00 a	10.47 ± 2.39 b	9.14 ± 2.26 b	7.90 ± 0.50 b	5.52 ± 0.99 b,c	1.57 ± 0.71 c
Bulk density (g/cm^3)	0.15 ± 0.00 c	0.16 ± 0.00 b,c	0.17 ± 0.00 b	0.17 ± 0.00 b	0.18 ± 0.00 b	0.29 ± 0.01 a
Available water holding capacity (% *v*/*v*)	66.54 ± 1.75 a	62.21 ± 1.28 a	68.04 ± 2.57 a	45.41 ± 1.84 b	43.07 ± 3.00 b	68.30 ± 3.17 a

Total porosity (TP), available water holding capacity (AWHC), air-filled porosity (AFP), bulk density (BD) by volume. Values are mean± SE (n = 4). In each row, values followed by the same letter do not differ significantly, according to Duncan's multiple range test at $p < 0.05$.

Table 2. Growing media (peat, *Sideritis cypria* waste—SCW) physicochemical properties before plant transplanting.

	Peat 100%	SCW 5%	SCW 10%	SCW 20%	SCW 40%	SCW 100%
pH	6.32 ± 1.79 d	6.27 ± 1.12 d	6.53 ± 0.00 c,d	6.91 ± 0.04 b	7.54 ± 0.01 a	6.75 ± 0.08 b,c
EC (mS/cm)	0.84 ± 0.04 b	1.26 ± 0.07 a	1.24 ± 0.11 a	1.10 ± 0.06 a	1.18 ± 0.03 a	1.19 ± 0.14 a
Organic matter (%)	72.38 ± 1.28 b	77.64 ± 2.26 b	75.39 ± 2.64 b	75.23 ± 1.45 b	76.11 ± 1.22 b	92.79 ± 0.25 a
Organic C (%)	41.98 ± 0.74 b	45.03 ± 1.31 b	43.72 ± 1.53 b	43.63 ± 0.84 b	44.14 ± 0.71 b	53.82 ± 0.14 a
C/N ratio	50.36 ± 2.12 a	50.89 ± 1.44 a	38.38 ± 1.21 b,c	36.33 ± 1.54 c	30.04 ± 0.88 d	42.57 ± 1.19 b
N (g/kg)	8.35 ± 0.19 c	8.87 ± 0.45 c	11.40 ± 0.46 b	12.04 ± 0.44 b	14.73 ± 0.65 a	12.66 ± 0.37 b
K (g/kg)	2.03 ± 0.05 f	3.11 ± 0.25 e	4.20 ± 0.32 d	6.31 ± 0.25 c	8.75 ± 0.25 b	14.06 ± 0.35 a
P (g/kg)	1.12 ± 0.04 c	1.39 ± 0.05 bc	1.70 ± 0.06 b	1.75 ± 0.07 b	2.51 ± 0.28 a	1.65 ± 0.10 b
Ca (g/kg)	15.01 ± 0.56 b	16.37 ± 1.19 b	20.02 ± 0.33 a	22.57 ± 1.07 a	21.27 ± 0.73 a	11.58 ± 0.50 c
Mg (g/kg)	0.79 ± 0.04 d	1.02 ± 0.08 d	1.30 ± 0.03 c	1.82 ± 0.08 b	2.25 ± 0.06 a	1.70 ± 0.11 b
Na (g/kg)	0.97 ± 0.03 e	1.03 ± 0.07 e	1.20 ± 0.05 d	1.65 ± 0.03 c	2.03 ± 0.01 b	5.79 ± 0.04 a
Total porosity %	84.97 ± 0.76 a,b	91.82 ± 3.87 a	77.63 ± 3.41 b,c	69.01 ± 3.70 c,d	62.88 ± 5.28 d	98.19 ± 1.62 a
Air-filled porosity (% *v*/*v*)	18.43 ± 1.00 a	15.52 ± 2.21 a,b	14.28 ± 1.46 a,b	13.42 ± 2.65 a,b	9.62 ± 2.91 b	17.14 ± 1.43 a,b
Bulk density (g/cm^3)	0.15 ± 0.00 b	0.16 ± 0.00 a	0.17 ± 0.00 a	0.17 ± 0.00 a	0.15 ± 0.00 b	0.12 ± 0.00 c
Available water holding capacity (% *v*/*v*)	66.54 ± 1.75 b	76.30 ± 1.68 a	63.35 ± 2.56 b	55.57 ± 1.44 c	53.25 ± 2.37 c	81.05 ± 0.19 a

Total porosity (TP), available water holding capacity (AWHC), air-filled porosity (AFP), bulk density (BD) by volume. Values are mean± SE (n = 4). In each row, values followed by the same letter do not differ significantly, according to Duncan's multiple range test at $p < 0.05$.

According to Zhou et al. [55], herbal residues may contain cellulose, protein, and polysaccharides that can provide soil with N, P, and K after the organic matter decomposition [56]. This was also evidenced in the MAP-enriched growing media in the present study, as the N, K, Mg, and P levels were significantly increased with increasing ratios of waste in the media. On the other hand, the EC value of the growing media was higher (ranging from 0.84 to 1.70 mS/cm) than the suggested ones for growing media [17,53]. However, any mineral amendments can always been regulated by utilizing a tailor-made fertilization strategy or by using inert materials (e.g., perlite) in case of excessive mineral contents that may increase the EC value of the growing media [57].

It has been reported that the addition of mint distillation wastes increased mustard (*Brassica juncea*) productivity and also improved soil physicochemical properties, making it possible to partly substitute the fertilizer inputs [58]. Particle size, overall porosity, air-filled capacity, and water holding capacity of the growing medium are critical to the success of soil-less culture crops. In both ODW- and SCW-based growing media, total porosity (especially at high waste ratios) decreased, which resulted in decreased air-filled capacity (up to 5.52%) and available water holding capacity (up to 43.07%), suggesting an overall negative effect on the media's physical properties. Moreover, the total pore space of the tested mixtures was lower than the recommended values of ≥85% porosity (except at SCW 5%), according to Abad et al. [53]. Therefore, inert materials with big particle size, such as perlite, could be added to the growing medium to increase the porosity and alleviate the negative effects of MAP waste on this particular parameter. Another important parameter is air capacity (air-filled porosity) with recommended values of 20–30% for growing media [53]. In our study, decreasing trends for this parameter were recorded for both waste materials, especially in the case of ODW, where a gradual decrease was recorded with increasing waste ratios in the growing media. On the other hand, SCW performed better with the lowest values being recorded for the SCW 40% treatment, while lower rates of SCW did not differ significantly from peat despite the decreasing trends with increasing amounts of SCW.

The effects of adding MAP waste in peat-based growing media on purslane growing parameters are presented in Table 3. The presence of ODW or SCW in the growing media at levels ≥ 10% reduced plant height and the number of leaves produced, with more noticeable impacts at the highest SCW ratio (SCW 40%) and the treatments of ODW 20% and 40%, in comparison with control treatment (peat 100%). This negative impact on plant growth also derived in decreased fresh biomass, even at ODW 5% or SCW 10%, as well as in decreased dry weight, especially in the case of ODW ratios ≥ 20% and SCW 40% treatment (Table 3). In other studies, the addition of MAP residues from Chinese herbs in the soil boosted the dry matter content of tomato and cabbage plants [8], whereas in the present experiment, dry weight was similar or lower than control (peat). Similarly, the addition of olive mill waste in peat at high ratios (>10%) could reduce the growth of potted ornamental plants and the market potential of the harvested product [24]. In contrast, composted olive mill waste increased shoot biomass when applied on tomato crop [59], highlighting the benefits obtained of a stable material, derived through the composting process instead of using raw waste material. Indeed, it is simpler to handle, store, and transfer smaller plants, since the reduction in plant height in nurseries or greenhouses is not always a negative parameter [24,60]. These findings demonstrate the variable effects of introducing crop residues to soil- and soil-less-culture systems. However, the use of high ratios of MAP residues employed in the current study should be excluded unless other crop management practices that improve plant productivity could be considered to mitigate the lower fresh biomass production, which consequently affects crop productivity. These practices involve the employment of a fertigation system, the enhancement of the physicochemical characteristics of the growing medium, and possibly the use of semi- or fully composted material. Reduced plant growth could be mostly associated with substrate parameters, such as total and air-filled porosity. Therefore, proper aeration of the growing medium could be enhanced by incorporating inert materials in the mixture, such as perlite, pumice, or sand, up to 15–20%, resulting in improved results for plant growth and development.

Leaf stomatal conductance was decreased in purslane plants grown in ≥10% ODW and ≥5% SCW when compared with the control treatment (peat 100%), with greater effects being evident at the highest ratio (e.g., 40%) of the wastes into the media (Table 4). These results indicate that increasing the waste ratio at values higher than 5% and 10% for SCW and ODW puts purslane plants under stressful conditions since stomatal closure is associated with a plant's response to stress. The results of the current study are consistent with earlier studies on stomatal closure in tomato plants grown in sand and irrigated with olive mill wastewater [61], or in *Brassica* seedlings grown in peat-based media with olive mill waste [15]. Therefore, it could be suggested that the incorporation of waste material in growing media should be carefully considered, since high EC values may impose waster stress on plants due to the high osmotic potential of the growing media.

Table 3. Impact of growing media (peat; *Origanum dubium* waste—ODW; *Sideritis cypria* waste—SCW) on purslane seedlings' height (cm), leaf number, upper part fresh weight (g/plant), and dry weight (g/plant) on plants grown in greenhouse/nursery.

	Height	Leaf No.	Fresh Weight	Dry Weight
Peat 100%	22.74 ± 0.79 a [Y]	41.00 ± 3.11 a	12.52 ± 0.60 a	0.56 ± 0.01 a
ODW 5%	18.12 ± 3.31 a,b	27.60 ±5.80 a,b	6.43 ± 2.30 b	0.54 ± 0.20 a
ODW 10%	15.48 ± 3.03 b	21.60 ±8.50 b	6.28 ± 2.76 b	0.39 ± 0.17 a,b
ODW 20%	9.34 ± 0.66 c	7.60 ± 0.74 c	0.98 ± 0.12 c	0.07 ± 0.01 b
ODW 40%	7.30 ± 0.99 c	4.00 ± 0.71 c	0.34 ± 0.09 c	0.03 ± 0.00 b
Peat 100%	22.74 ± 0.79 a	41.00 ± 3.11 a	12.52 ± 0.60 a	0.56 ± 0.01 a
SCW 5%	22.48 ± 1.74 a	39.40 ± 3.04 a	10.36 ± 1.64 a,b	0.46 ± 0.11 a
SCW 10%	17.04 ± 2.77 b	27.60 ± 8.80 a,b	5.55 ± 2.27 bc	0.49 ± 0.16 a
SCW 20%	15.90 ± 1.79 b,c	21.00 ± 3.17 b,c	3.93 ± 0.67 c	0.30 ± 0.11 a,b
SCW 40%	11.24 ± 1.56 c	10.80 ± 0.96 c	1.51 ± 0.32 c	0.11 ± 0.01 b

[Y] Mean values (n = 6) in the same column followed by the same letter for the same waste material (ODW or SCW), are not significantly different according to Duncan's multiple range test (DMRT), at $p < 0.05$.

Table 4. Impact of growing media (peat, *Origanum dubium* waste—ODW, *Sideritis cypria* waste—SCW) on purslane chlorophyll fluorescence (Fv/Fm), stomatal conductance (s/cm), chlorophylls (Chl a, Chl b, total Chls; mg/g fw), and carotenoid (mg/g fw) content.

	Stomatal Conductance	Chlorophyll Fluorescence	Chl a	Chl b	Total Chls	Carotenoids
Peat 100%	110.16 ± 8.67 a [Y]	0.80 ± 0.00 a	0.39 ± 0.02 a	0.10 ± 0.00 a	0.49 ± 0.03 a	0.07 ± 0.00 a
ODW 5%	82.00 ± 5.50 a,b	0.77 ± 0.00 b	0.36 ± 0.06 a	0.08 ± 0.01 a	0.45 ± 0.07 a	0.08 ± 0.01 a
ODW 10%	65.00 ± 5.85 b	0.75 ± 0.00 b,c	0.20 ±0.03 b	0.04 ± 0.01 b	0.24 ± 0.04 b	0.04 ± 0.00 b
ODW 20%	62.5 ± 0.50 b	0.74 ± 0.00 c	0.18 ±0.04 b	0.03 ± 0.01 b,c	0.21 ± 0.05 b	0.04 ± 0.01 b
ODW 40%	n.m.	n.m.	0.09 ±0.00 b	0.01 ± 0.00 c	0.09 ± 0.00 b	0.03 ± 0.00 b
Peat 100%	110.16 ± 8.67 a	0.80 ± 0.00 a	0.39 ± 0.02 a	0.10 ± 0.00 a	0.49 ± 0.03 a	0.07 ± 0.00 a
SCW 5%	69.00 ± 6.80 b	0.72 ± 0.03 b	0.28 ± 0.03 b	0.06 ± 0.01 b	0.35 ± 0.04 b	0.05 ± 0.01 b
SCW 10%	59.50 ± 3.50 b,c	0.75 ± 0.01 a,b	0.18 ± 0.00 c,d	0.04 ± 0.00 c,d	0.22 ± 0.00 c,d	0.04 ± 0.00 b,c
SCW 20%	55.00 ± 5.85 b,c	0.74 ± 0.01 a,b	0.24 ± 0.01 b,c	0.05 ± 0.00 b,c	0.29 ± 0.01 b,c	0.05 ± 0.00 b,c
SCW 40%	33.50 ± 0.50 c	0.65 ± 0.01 c	0.14 ± 0.00 d	0.03 ± 0.00 d	0.17 ± 0.00 d	0.03 ± 0.00 c

[Y] Mean values (n = 6) in columns followed by the same letter for the different wastes (ODW, SCW) are not significantly different, according to Duncan's multiple range test (DMRT), at $p < 0.05$. n.m.: not measured due to small leaf area.

The synthesis of leaf pigments, such as chlorophyll a, chlorophyll b, and total chlorophylls, as well as carotenoids, was decreased in purslane plants grown in ≥10% ODW and ≥10% SCW media, compared with purslane grown in 100% peat (Table 4). Chlorophyll content is directly related to the photosynthetic capacity of the plant [62]. Therefore, the reduced chlorophyll levels recorded in the present work reflect the reduced plant growth when high ODW and SCW ratios were implemented. MAP residues improved mineral availability to the plants through the increased mineral content of the growing media mixtures; however, the use of raw residues may have consumed a portion of the available N for the decomposition process, on one hand, whereas the negative effects on the physicochemical properties of the mixtures, on the other hand, could have prevented the efficient supply of minerals to the plants. Therefore, the particular decrease in plant growth in the present study should not be related to the mineral status of the growing media. Instead, it is more likely related to the negative effects on the physicochemical properties of the growing media, such as air-filled porosity. In this regard, initiatives to improve the qualities of the growing media should be considered, either increasing the quantity of inert material (i.e., adding 15–20% of perlite) or combining different inert materials (perlite, sand, zeolite, vermiculate, etc.) along with the tested waste. Moreover, efforts should be made to supplement N to counteract N losses due to its utilization by microorganisms for organic matter decomposition.

Table 5 presents the content of minerals accumulated in purslane plants grown in ODW- or SCW-based growing media after a cultivation period of 25 days. Plants grown in ODW-based media revealed lower N, Na, and Mg accumulation compared with plants grown in peat. However, the lowest N and Na content were found in ODW mixtures that contained 5–10% of ODW, indicating the contribution of ODW to mineral accumulation in the growing media. The highest Mg content was found in plants grown in peat, while the highest Ca contest was found in plants grown in ODW 40% media. In the case of SCW mixtures, N and Mg levels decreased with the presence of SCW in the growing media, whereas Na content increased in plants grown in both control (100% peat) and SCW 40%. No main differences were found in Ca content in SCW-based grown plants.

Table 5. Impact of growing media (peat, *Origanum dubium* waste—ODW, *Sideritis cypria* waste—SCW) on mineral element content (mg/g dry weight) of purslane plants.

	N	K	P	Na	Ca	Mg
Peat 100%	35.22 ± 0.27 a [Y]	86.96 ± 1.10 b	11.40 ± 0.31 a,b	12.12 ± 0.09 c	6.26 ± 0.16 a	5.20 ± 0.18 a
ODW 5%	27.87 ± 1.09 c	84.13 ± 4.11 b	9.27 ± 1.56 b,c	11.78 ± 0.56 c	5.56 ± 0.05 b,c	3.14 ± 0.46 b
ODW 10%	25.34 ± 0.48 d	99.95 ± 4.05 a	12.17 ± 0.59 a	12.41 ± 0.22 c	5.33 ± 0.30 c	3.29 ± 0.40 b
ODW 20%	32.74 ± 0.32 b	86.31 ± 1.69 b	10.92 ± 0.21 a,b	14.61 ± 0.28 b	6.10 ± 0.12 a,b	3.61 ± 0.07 b
ODW 40%	31.23 ± 0.24 b	60.35 ± 1.18 c	7.15 ± 0.14 c	16.96 ± 0.33 a	5.40 ± 1.10 c	2.81 ± 0.05 b
Peat 100%	35.22 ± 0.27 a	86.96 ± 1.10 b,c	11.40 ± 0.31 b	12.12 ± 0.09 b	6.26 ± 0.16 a	5.20 ± 0.18 a
SCW 5%	28.56 ± 0.15 b	84.55 ± 0.54 b,c	8.51 ± 0.64 c	10.71 ± 0.05 c	5.60 ± 0.04 b	3.96 ± 0.06 b
SCW 10%	25.95 ± 0.87 c	89.59 ± 3.90 b	11.51 ± 0.27 b	11.42 ± 0.33 c	6.15 ± 0.21 a	3.92 ± 0.28 b
SCW 20%	25.92 ± 0.12 c	100.43 ± 1.96 a	12.85 ± 0.25 a	11.17 ± 0.22 c	4.99 ± 0.09 c	3.55 ± 0.07 b
SCW 40%	25.89 ± 0.14 c	81.78 ± 1.60 c	5.49 ± 0.11 d	13.69 ± 0.26 a	4.86 ± 0.09 c	4.05 ± 0.08 b

[Y] Mean values (n = 4) in columns followed by the same letter for the different wastes (ODW, SCW) are not significantly different, according to Duncan's multiple range test (DMRT), at $p < 0.05$.

Our results indicate that when ODW or SCW were added, the N that was provided was mostly organic and only partially available to the plants. Moreover, a large part of N seems to be consumed by microorganisms through organic matter decomposition, as illustrated by the decreased C/N ratios (Tables 1 and 2), the decreased N accumulation in purslane tissues (Table 5), and the decreased N levels at the growing media after harvesting (Tables S1 and S2). This finding could explain why plants grew slowly and had limited quantities of chlorophyll and photosynthetic capacity, parameters that are directly associated with N availability [61]. One of the biggest drawbacks of using raw materials in growing media is the variability in the nutrients that are accumulated in plant tissues. Another issue is the practical challenges of performing preliminary studies prior to the usage of MAP-based media. This can be prevented by implementing a tailor-made fertilization regime [63], since, according Chrysargyris et al. [24], the use of an adjusted hydroponic fertilizer solution in ornamental plants may alleviate mineral abnormalities caused by the uncomposted growing medium.

The use of MAP waste as a partial substitute of peat into plant growing media also affected the total phenolic compound content, total flavonoids, and antioxidant capacity of the produced plants, as presented in Figure 1. In the case of ODW, the content of total phenolic compounds in purslane was increased up to 30.2% for ratios of ODW between 5% and 20% and further increased 1.5 times at the highest ratio of ODW (ODW 40%) compared with the control treatment (Figure 1(A1)). Purslane antioxidant activity, as assayed by DPPH and FRAP, significantly increased at ≥20% of ODW and ≥5% of SCW, respectively, while the highest ratio of ODW (ODW 40%) more than doubled the antioxidant activity for both assays compared with the control treatment (Figure 1(A2,A3,B2,B3)). Similarly, total flavonoid content increased (up to 2.2 times) at 40% ODW in comparison with the control treatment, while a less profound increase was recorded for the respective treatment of SCW (Figure 1(A4,B4)). The main response of plants to stress conditions is reflected by the activation of several enzymatic and nonenzymatic mechanisms. In this case, various nonenzymatic responses were indicated by increased phenolic compounds, flavonoids, and antioxidant capacity of purslane. Therefore, despite the decreased yield observed when high ratios of ODW and SCW were implemented, the improved antioxidant capacity of plants could be important since it is associated with an increased nutritional value of the plant (high antioxidant compound content). Similar responses of *Brassica* plants to stress conditions when they were grown in OMW-based media were also reported, confirming our argument that the incorporation of raw crop residues in growing media may put plants under stressful conditions and consequently result in increased antioxidant compound content [15].

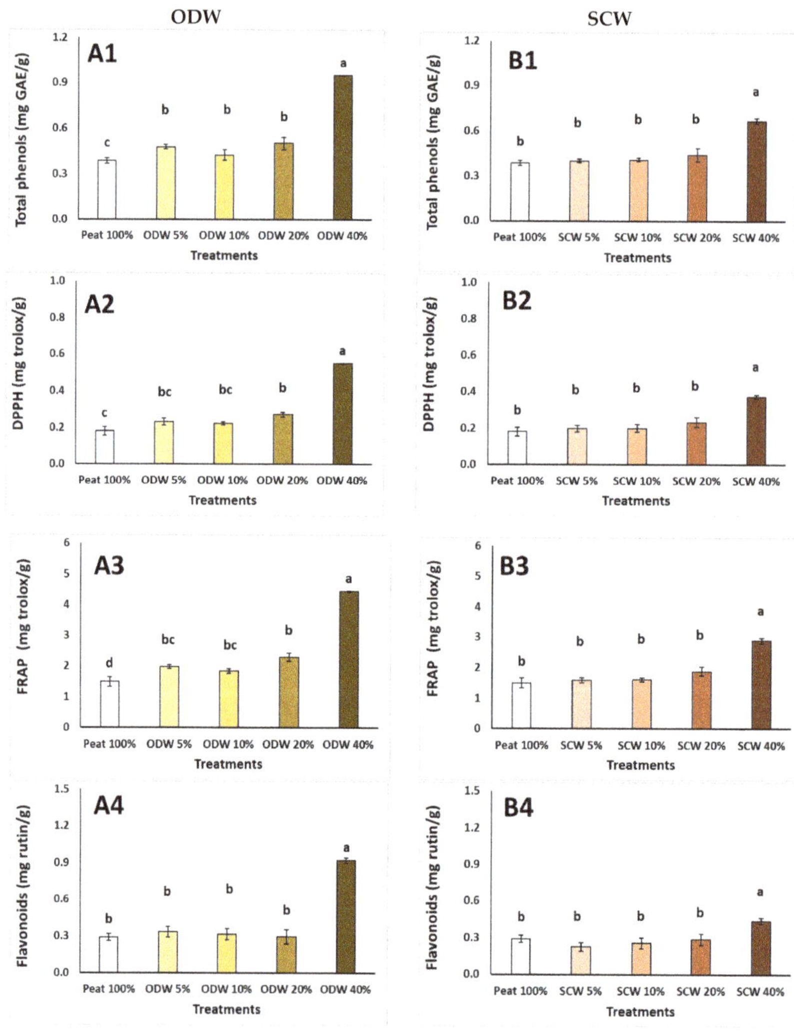

Figure 1. Effect of growing media (peat, *Origanum dubium* waste—ODW, *Sideritis cypria* waste—SCW) on total phenolic compound content (mg GAE/g fw), antioxidant activity (DPPH, FRAP: mg Trolox/g fw), and flavonoid content (mg rutin/g fw) of purslane plants (ODW: **A1–A4**; SCW: **B1–B4**). Values are means± SE (n = 6). Mean values followed by the same letter do not differ significantly at $p \geq 0.05$ according to Duncan's multiple range test.

Throughout their growth cycle, plants are exposed to a range of challenging stress situations. These situations frequently entail both biotic (such as diseases and pests) and abiotic stressors, such as intense heat, dry seasons, extremely saline and osmotic soils, and high mineral/heavy metal concentrations. Plants have developed a number of detoxification processes to overcome the oxidative stress from the reactive oxygen species that accumulate in cells under stressful conditions. The formation of MDA, which is linked to increased lipid peroxidation under stress conditions, is one of the most used stress markers. MDA levels increase when plant antioxidants fail to scavenge ROS as a first-step response to stressors. In our study, MDA content increased in purslane plants grown at ODW 40%, having also high levels of H_2O_2, which indicate stress conditions associated with cellular damage (Table 6). This was evidenced by the increased SOD activity, followed by increased POD activity for the same treatment (ODW 40%). Several nonenzymatic (phenols, proline, ascorbic acid, etc.) and enzymatic antioxidants (SOD, CAT, POD, etc.) are activated to neutralize ROS accumulation in cells [64]. In the case of SCW waste, MDA levels were low at the high SCW ratios (20–40%), and H_2O_2 and POD levels increased at the highest SCW ratio (SCW 40%), whereas SOD levels were not affected by the ratio of SCW in the growing medium. This indicates that the antioxidant role of SOD was exhausted to decreased MDA levels, and the POD increased in order to scavenge the high levels of H_2O_2.

Table 6. Impact of growing media (peat, *Origanum dubium* waste—ODW, *Sideritis cypria* waste—SCW) on hydrogen peroxide (H_2O_2) (μmol/g), lipid peroxidation (MDA) (nmol/g), and antioxidant enzyme activity of superoxide dismutase (SOD; units/mg protein), catalase (CAT; units/mg protein), and peroxidase (POD; units/mg protein) of purslane plants.

	H_2O_2	MDA	SOD	CAT	POD
Peat 100%	0.20 ± 0.01 b,c [Y]	8.84 ± 0.74 b	0.94 ± 0.12 b	3.61 ± 0.71 b,c	0.54 ± 0.05 c
ODW 5%	0.23 ± 0.01 b,c	8.68 ± 0.59 b,c	1.15 ± 0.07 b	11.65 ± 0.85 a	0.81 ± 0.06 b
ODW 10%	0.19 ± 0.00 c	7.46 ± 0.86 c	0.82 ± 0.11 b	5.10 ± 0.60 b	0.68 ± 0.03 b,c
ODW 20%	0.24 ± 0.01 b	10.93 ± 0.91 a,b	1.08 ± 0.08 b	1.68 ± 0.41 c	0.86 ± 0.09 b
ODW 40%	0.46 ± 0.03 a	12.87 ± 0.71 a	1.85 ± 0.04 a	3.19 ± 0.35 b,c	1.30 ± 0.16 a
Peat 100%	0.15 ± 0.00 b,c	7.85 ± 0.57 a	0.94 ± 0.12	3.61 ± 0.71 b,c	0.54 ± 0.05 c
SCW 5%	0.17 ± 0.01 a,b	5.25 ± 0.67 b	1.13 ± 0.04	4.69 ± 0.22 b	0.86 ± 0.03 a,b
SCW 10%	0.12 ± 0.00 c	5.11 ± 0.78 b,c	1.01 ± 0.12	9.03 ± 0.91 a	0.84 ± 0.07 a,b
SCW 20%	0.15 ± 0.02 b,c	3.73 ± 0.14 b,c	1.14 ± 0.07	2.04 ± 0.58 c	0.76 ± 0.02 b
SCW 40%	0.21 ± 0.03 a	3.27 ± 0.52 c	1.15 ± 0.07	2.38 ± 0.85 b,c	0.99 ± 0.08 a

[Y] Values ($n = 4$) in columns followed by the same letter for the different wastes (ODW, SCW) are not significantly different, according to Duncan's multiple range test at $p < 0.05$.

For commercially producing vegetables, ornamentals, seedlings, and potted plants, it is essential to ensure the rapid growth and vigor of plants. In addition, it is crucial to use peat substitutes that are more environmentally friendly by mixing them in growing media in different combinations. Reduced peat use and consumption could lower seedling costs and contribute to the preservation of peatlands. Future research should focus on adjusting the ratios of the ingredients used to prepare commercial substrates, always considering the fertilization and irrigation regimes. For this purpose, minerals and organic matter found in crop residues can improve the physicochemical properties of the soil and growing medium. Even though some studies with potted plants may result in improved soil properties by adding MAP residues, there are still some differences between pot experiments and actual agricultural practice in order to extrapolate these results to field condition and suggest the introduction of MAP waste in commercial farming [10].

4. Conclusions

The two most important issues that researchers should focus on when evaluating new materials as potential components in substrate mixtures are to increase growing media fertility and to establish the appropriate physicochemical properties of the final mixture. In the current study, ODW and SCW were explored as a partial substitute for peat in the growth of purslane seedlings. The physicochemical properties of the growing media were altered by the addition of both MAP residues to the peat-based mixtures. The media's ability to hold water and free air was significantly reduced as a consequence of the observed reduction in free pore space. Nonetheless, there was an increase in the amount of organic matter and minerals that were available for use by plants in both of the used wastes. The addition of ODW and SCW in the growing medium had a negative effect on a plant's physiological characteristics and reduced leaf stomatal conductance, especially at high residue ratios of 20–40%. Moreover, the degradation of organic waste increased mineral availability, which accumulated in plant tissues (especially P, K, and Na). However, the use of high ratios of MAP waste significantly inhibited the growth of purslane plants, a finding that could be attributed to the development of stress conditions since both nonenzymatic and enzymatic antioxidant mechanisms were induced. The current study concludes that both ODW and SCW can be used in growth media at low ratios (up to 10%), which, despite a slight growth inhibition, may result in increased antioxidant compound content and improved nutritional benefits of the edible plant parts. However, the introduction of these new materials as peat substitutes requires the investigation of agronomic practices that could further improve the properties of the growing media's improvement and ensure high crop yields. MAP residues can also be considered to fertilize agricultural soils that can soften the observed negative effects found on the growing media's properties. Additionally, the use of inert materials with big particle size, such as perlite, in the growing media in higher ratios, i.e., 20–30%, could increase porosity and alleviate the negative effects of MAP waste on this particular parameter. Composting is also a practice that can be considered in order to produce stable material, but limitations of the process also need to be considered, such as time, space, energy, and cost-demanding process.

Supplementary Materials: The following supporting information can be downloaded at: https://www.mdpi.com/article/10.3390/horticulturae9040474/s1, Figure S1. MAP residues' particle size and uniformity index evaluation for *Origanum dubium* waste (ODW) and *Sideritis cypria* waste (SCW). Table S1. Growing media's (peat, *Origanum dubium* waste (ODW)) physicochemical properties after plant harvesting (at the end of the growing period). In parenthesis is the percentage of properties' changes compared with the start of the experiment—see Table 1. Table S2. Growing media's (peat, *Sideritis cypria* waste (SCW)) physicochemical properties after plant harvesting (at the end of the growing period). In parenthesis is the percentage of properties' changes compared with the start of the experiment—see Table 2.

Author Contributions: Conceptualization, A.C. and N.T.; methodology, A.C., S.A.P. and N.T.; software, A.C.; validation, A.C. and N.T.; formal analysis, A.C. and N.T.; investigation, A.C., S.L. and N.T.; resources, N.T.; data curation, A.C. and S.L.; writing—original draft preparation, A.C. and N.T.; writing—review and editing, A.C., S.A.P. and N.T.; visualization, A.C., S.A.P. and N.T.; supervision, A.C. and N.T.; project administration, N.T.; funding acquisition, N.T. All authors have read and agreed to the published version of the manuscript.

Funding: This research was funded by PRIMA (ValueFarm project) (grant Number Prima2019-11, PRIMA/0009/2019, P2P/PRIMA/1218/0006, 01DH20006, Prima2019-12, STDF Valuefarm, 18-3-2021, TUBITAK-119N494, 301 / October 18th, 2020, PCI2020-112091) a programme supported by the European Union with co-funding by the Funding Agencies RIF—Cyprus.

Institutional Review Board Statement: Not applicable.

Data Availability Statement: Not applicable.

Conflicts of Interest: The authors declare no conflict of interest.

References

1. Santana-Méridas, O.; González-Coloma, A.; Sánchez-Vioque, R. Agricultural residues as a source of bioactive natural products. *Phytochem. Rev.* **2012**, *11*, 447–466. [CrossRef]
2. Saha, A.; Basak, B.B. Scope of value addition and utilization of residual biomass from medicinal and aromatic plants. *Ind. Crops Prod.* **2020**, *145*, 111979. [CrossRef]
3. Olofsson, J.; Börjesson, P. Residual biomass as resource—Life-cycle environmental impact of wastes in circular resource systems. *J. Clean. Prod.* **2018**, *196*, 997–1006. [CrossRef]
4. Ferhat, M.A.; Meklati, B.Y.; Chemat, F. Comparison of different isolation methods of essential oil from Citrus fruits: Cold pressing, hydrodistillation and microwave 'dry' distillation Mohamed. *Flavour Fragr. J.* **2008**, *22*, 494–504. [CrossRef]
5. Saha, A.; Tripathy, V.; Basak, B.B.; Kumar, J. Entrapment of distilled palmarosa (*Cymbopogon martinii*) wastes in alginate beads for adsorptive removal of methylene blue from aqueous solution. *Environ. Prog. Sustain. Energy* **2018**, *37*, 1942–1953. [CrossRef]
6. Sayed-Ahmad, B.; Talou, T.; Saad, Z.; Hijazi, A.; Merah, O. The Apiaceae: Ethnomedicinal family as source for industrial uses. *Ind. Crops Prod.* **2017**, *109*, 661–671. [CrossRef]
7. Skendi, A.; Irakli, M.; Chatzopoulou, P.; Bouloumpasi, E.; Biliaderis, C.G. Phenolic extracts from solid wastes of the aromatic plant essential oil industry: Potential uses in food applications. *Food Chem. Adv.* **2022**, *1*, 100065. [CrossRef]
8. Zhou, Y.; Manu, M.K.; Li, D.; Johnravindar, D.; Selvam, A.; Varjani, S.; Wong, J. Effect of Chinese medicinal herbal residues compost on tomato and Chinese cabbage plants: Assessment on phytopathogenic effect and nutrients uptake. *Environ. Res.* **2023**, *216*, 114747. [CrossRef] [PubMed]
9. Gruda, N.S. Increasing sustainability of growing media constituents and stand-alone substrates in soilless culture systems. *Agronomy* **2019**, *9*, 298. [CrossRef]
10. Ma, J.; Chen, Y.; Wang, K.; Huang, Y.; Wang, H. Re-utilization of Chinese medicinal herbal residues improved soil fertility and maintained maize yield under chemical fertilizer reduction. *Chemosphere* **2021**, *283*, 131262. [CrossRef] [PubMed]
11. Ratiarisoa, R.V.; Magniont, C.; Ginestet, S.; Oms, C.; Escadeillas, G. Assessment of distilled lavender stalks as bioaggregate for building materials: Hygrothermal properties, mechanical performance and chemical interactions with mineral pozzolanic binder. *Constr. Build. Mater.* **2016**, *124*, 801–815. [CrossRef]
12. Singh, P.; Hundal, J.S.; Patra, A.K.; Wadhwa, M.; Sharma, A. Sustainable utilization of *Aloe vera* waste in the diet of lactating cows for improvement of milk production performance and reduction of carbon footprint. *J. Clean. Prod.* **2021**, *288*, 125118. [CrossRef]
13. Xiao, D.; Shao, H.; Huo, Y.; Agung Nugroho, W.; Ifeoluwa Ogunniran, B.; Fan, W.; Huo, M. Reclamation of ginseng residues using two-stage fermentation and evaluation of their beneficial effects as dietary feed supplements for piglets. *Waste Manag.* **2022**, *154*, 293–302. [CrossRef] [PubMed]
14. Chrysargyris, A.; Antoniou, O.; Xylia, P.; Petropoulos, S.; Tzortzakis, N. The use of spent coffee grounds in growing media for the production of *Brassica* seedlings in nurseries. *Environ. Sci. Pollut. Res.* **2021**, *28*, 24279–24290. [CrossRef] [PubMed]
15. Chrysargyris, A.; Antoniou, O.; Athinodorou, F.; Vassiliou, R.; Papadaki, A.; Tzortzakis, N. Deployment of olive-stone waste as a substitute growing medium component for *Brassica* seedling production in nurseries. *Environ. Sci. Pollut. Res.* **2019**, *26*, 35461–35472. [CrossRef] [PubMed]
16. Patsalou, M.; Chrysargyris, A.; Tzortzakis, N.; Koutinas, M. A biorefinery for conversion of citrus peel waste into essential oils, pectin, fertilizer and succinic acid via different fermentation strategies. *Waste Manag.* **2020**, *113*, 469–477. [CrossRef] [PubMed]
17. Bolechowski, A.; Moral, R.; Bustamante, M.A.; Bartual, J.; Paredes, C.; Pérez-Murcia, M.D.; Carbonell-Barrachina, A.A. Winery-distillery composts as partial substitutes of traditional growing media: Effect on the volatile composition of thyme essential oils. *Sci. Hortic.* **2015**, *193*, 69–76. [CrossRef]
18. Lu, Q.; Li, C. Comprehensive utilization of Chinese medicine residues for industry and environment protection: Turning waste into treasure. *J. Clean. Prod.* **2021**, *279*, 123856. [CrossRef]
19. Ilangovan, M.; Guna, V.; Hu, C.; Nagananda, G.S.; Reddy, N. *Curcuma longa* L. plant residue as a source for natural cellulose fibers with antimicrobial activity. *Ind. Crops Prod.* **2018**, *112*, 556–560. [CrossRef]
20. Bistgani, Z.E.; Hashemi, M.; DaCosta, M.; Craker, L.; Maggi, F.; Morshedloo, M.R. Effect of salinity stress on the physiological characteristics, phenolic compounds and antioxidant activity of *Thymus vulgaris* L. and *Thymus daenensis* Celak. *Ind. Crops Prod.* **2019**, *135*, 311–320. [CrossRef]
21. Bustamante, M.A.; Paredes, C.; Moral, R.; Agulló, E.; Pérez-Murcia, M.D.; Abad, M. Composts from distillery wastes as peat substitutes for transplant production. *Resour. Conserv. Recycl.* **2008**, *52*, 792–799. [CrossRef]
22. Giménez, A.; Fernández, J.A.; Pascual, J.A.; Ros, M.; Saez-Tovar, J.; Martinez-Sabater, E.; Gruda, N.S.; Egea-Gilabert, C. Promising composts as growing media for the production of baby leaf lettuce in a floating system. *Agronomy* **2020**, *10*, 1540. [CrossRef]
23. Ceglie, F.G.; Bustamante, M.A.; Ben Amara, M.; Tittarelli, F. The challenge of peat substitution in organic seedling production: Optimization of growing media formulation through mixture design and response surface analysis. *PLoS ONE* **2015**, *10*, e0128600. [CrossRef]
24. Chrysargyris, A.; Antoniou, O.; Tzionis, A.; Prasad, M.; Tzortzakis, N. Alternative soilless media using olive-mill and paper waste for growing ornamental plants. *Environ. Sci. Pollut. Res.* **2018**, *25*, 35915–35927. [CrossRef]
25. Petropoulos, S.; Fernandes, Â.; Stojković, D.; Pereira, C.; Taofiq, O.; Di Gioia, F.; Tzortzakis, N.; Soković, M.; Barros, L.; Ferreira, I.C.F.R. Cotton and cardoon byproducts as potential growing media components for *Cichorium spinosum* L. commercial cultivation. *J. Clean. Prod.* **2019**, *240*, 118254. [CrossRef]

26. Chrysargyris, A.; Stamatakis, A.; Moustakas, K.; Prasad, M.; Tzortzakis, N. Evaluation of Municipal Solid Waste Compost and/or Fertigation as Peat Substituent for Pepper Seedlings Production. *Waste Biomass Valorization* **2018**, *9*, 2285–2294. [CrossRef]
27. Chrysargyris, A.; Tzortzakis, N. Municipal solid waste compost as a peat substitute for vegetable seedling production. A case study on cucumber and endive seedlings. In *Municipal Solid Waste: Management Strategies, Challenges and Future Directions*; Tzortzakis, N., Ed.; Nova Science Publishers, Inc.: Hauppauge, NY, USA, 2017; pp. 367–386.
28. Mill, L.A.; Najar, B.; Demasi, S.; Caser, M.; Gaino, W.; Cioni, P.L.; Pistelli, L.; Scariot, V. Cultivation substrate composition influences morphology, volatilome and essential oil of *Lavandula angustifolia* Mill. *Agronomy* **2019**, *9*, 411.
29. Fan, R.; Luo, J.; Yan, S.; Zhou, Y.; Zhang, Z. Effects of Biochar and Super Absorbent Polymer on Substrate Properties and Water Spinach Growth. *Pedosphere* **2015**, *25*, 737–748. [CrossRef]
30. Chrysargyris, A.; Saridakis, C.; Tzortzakis, N. Use of Municipal Solid Waste Compost as Growing Medium Component for Melon Seedlings Production. *J. Plant Biol. Soil Health* **2013**, *2*, 1–5.
31. Chrysargyris, A.; Prasad, M.; Kavanagh, A.; Tzortzakis, N. Biochar type, ratio, and nutrient levels in growing media affects seedling production and plant performance. *Agronomy* **2020**, *10*, 1421. [CrossRef]
32. Tubeileh, A.M.; Souikane, R.T. Effect of olive vegetation water and compost extracts on seed germination of four weed species. *Curr. Plant Biol.* **2020**, *22*, 100150. [CrossRef]
33. Kelepesi, S.; Tzortzakis, N.G. Olive mill wastesA growing medium component for seedling and crop production of lettuce and chicory. *Int. J. Veg. Sci.* **2009**, *15*, 325–339. [CrossRef]
34. Kuczmarski, D. Amending the cost of media. *Am. Nurs. Manag.* **1994**, *179*, 47–52.
35. Tsakaldimi, M. Kenaf (*Hibiscus cannabinus* L.) core and rice hulls as components of container media for growing Pinus halepensis M. seedlings. *Bioresour. Technol.* **2006**, *97*, 1631–1639. [CrossRef]
36. Chrysargyris, A.; Hajisolomou, E.; Xylia, P.; Tzortzakis, N. Olive-mill and grape-mill waste as a substitute growing media component for unexploded vegetables production. *Sustain. Chem. Pharm.* **2023**, *31*, 100940. [CrossRef]
37. Tebenkova, D.N.; Lukina, N.V.; Vorobyev, R.A.; Orlova, M.A.; Gagarin, Y.N. Germination and biometric parameters of seedlings grown on solid pulp and paper waste medium. *Contemp. Probl. Ecol.* **2015**, *8*, 892–900. [CrossRef]
38. Chugh, V.; Mishra, V.; Vishal Chugh, C.; Dwivedi, S.; Sharma, K. Purslane (*Portulaca oleracea* L.): An underutilized wonder plant with potential pharmacological value. *Pharma Innov. J.* **2019**, *8*, 236–246.
39. Kumar, A.; Sreedharan, S.; Kashyap, A.K.; Singh, P.; Ramchiary, N. A review on bioactive phytochemicals and ethnopharmacological potential of purslane (*Portulaca oleracea* L.). *Heliyon* **2022**, *8*, e08669. [CrossRef]
40. Petropoulos, S.A.; Fernandes, Â.; Calhelha, R.C.; Rouphael, Y.; Petrović, J.; Soković, M.; Ferreira, I.C.F.R.; Barros, L. Antimicrobial properties, cytotoxic effects, and fatty acids composition of vegetable oils from purslane, linseed, luffa, and pumpkin seeds. *Appl. Sci.* **2021**, *11*, 5738. [CrossRef]
41. Carrascosa, A.; Pascual, J.A.; Ros, M.; Petropoulos, S.A.; Alguacil, M. Agronomical Practices and Management for Commercial Cultivation of *Portulaca oleracea* as a Crop: A Review. *Plants* **2023**, *12*, 1246. [CrossRef]
42. Petropoulos, S.A.; Karkanis, A.; Martins, N.; Ferreira, I.C.F.R. Edible halophytes of the Mediterranean basin: Potential candidates for novel food products. *Trends Food Sci. Technol.* **2018**, *74*, 69–84. [CrossRef]
43. Thalassinos, G.; Petropoulos, S.A.; Antoniadis, V. The Response of Purslane (*Portulaca oleracea*) to Soil-Added Pb: Is It Suitable as a Potential Phytoremediation Species? *Toxics* **2023**, *11*, 153. [CrossRef]
44. Eliasson, J. *Sand/Media Specifications*; Washington State Department of Health: Tumwater, WA, USA, 2002.
45. *European Standard EN 13041*; Soil Improvers and Growing Media—Determination of Physical Properties—Dry Bulk Density, Air Volume, Water Volume, Shrinkage Value and Total Pore Space. European Committee for Standardization: Brussels, Belgium, 1999.
46. Chrysargyris, A.; Nikolaidou, E.; Stamatakis, A.; Tzortzakis, N. Vegetative, physiological, nutritional and antioxidant behavior of spearmint (*Mentha spicata* L.) in response to different nitrogen supply in hydroponics. *J. Appl. Res. Med. Aromat. Plants* **2017**, *6*, 52–61. [CrossRef]
47. Tzortzakis, N.; Pitsikoulaki, G.; Stamatakis, A.; Chrysargyris, A. Ammonium to Total Nitrogen Ratio Interactive Effects with Salinity Application on *Solanum lycopersicum* Growth, Physiology, and Fruit Storage in a Closed Hydroponic System. *Agronomy* **2022**, *12*, 386. [CrossRef]
48. Meyers, K.J.; Watkins, C.B.; Pritts, M.P.; Liu, R.H. Antioxidant and antiproliferative activities of strawberries. *J. Agric. Food Chem.* **2003**, *51*, 6887–6892. [CrossRef]
49. Loreto, F.; Velikova, V. Isoprene produced by leaves protects the photosynthetic apparatus against ozone damage, quenches ozone products, and reduces lipid peroxidation of cellular membranes. *Plant Physiol.* **2001**, *127*, 1781–1787. [CrossRef] [PubMed]
50. De Azevedo Neto, A.D.; Prisco, J.T.; Enéas-Filho, J.; De Abreu, C.E.B.; Gomes-Filho, E. Effect of salt stress on antioxidative enzymes and lipid peroxidation in leaves and roots of salt-tolerant and salt-sensitive maize genotypes. *Environ. Exp. Bot.* **2006**, *56*, 87–94. [CrossRef]
51. Rinaldi, S.; De Lucia, B.; Salvati, L.; Rea, E. Understanding complexity in the response of ornamental rosemary to different substrates: A multivariate analysis. *Sci. Hortic.* **2014**, *176*, 218–224. [CrossRef]
52. Chand, S.; Anwar, M.; Patra, D.D.; Khanuja, S.P.S. Effect of Mint Distillation Waste on Soil Microbial Biomass in a Mint-Mustard Cropping Sequence. *Commun. Soil Sci. Plant Anal.* **2004**, *35*, 243–254. [CrossRef]
53. Abad, M.; Noguera, P.; Burés, S. National inventory of organic wastes for use as growing media for ornamental potted plant production: Case study in Spain. *Bioresour. Technol.* **2001**, *77*, 197–200. [CrossRef]

54. Lasaridi, K.; Protopapa, I.; Kotsou, M.; Pilidis, G.; Manios, T.; Kyriacou, A. Quality assessment of composts in the Greek market: The need for standards and quality assurance. *J. Environ. Manag.* **2006**, *80*, 58–65. [CrossRef] [PubMed]
55. Zhou, Y.; Selvam, A.; Wong, J.W.C. Effect of Chinese medicinal herbal residues on microbial community succession and anti-pathogenic properties during co-composting with food waste. *Bioresour. Technol.* **2016**, *217*, 190–199. [CrossRef] [PubMed]
56. Wang, P.; Zhan, S.; Yu, H.; Xue, X.; Hong, N. The effects of temperature and catalysts on the pyrolysis of industrial wastes (herb residue). *Bioresour. Technol.* **2010**, *101*, 3236–3241. [CrossRef] [PubMed]
57. Ribeiro, H.M.; Romero, A.M.; Pereira, H.; Borges, P.; Cabral, F.; Vasconcelos, E. Evaluation of a compost obtained from forestry wastes and solid phase of pig slurry as a substrate for seedlings production. *Bioresour. Technol.* **2007**, *98*, 3294–3297. [CrossRef] [PubMed]
58. Patra, D.D.; Anwar, M.; Chand, S. Integrated nutrient management and waste recycling for restoring soil fertility and productivity in Japanese mint and mustard sequence in Uttar Pradesh, India. *Agric. Ecosyst. Environ.* **2000**, *80*, 267–275. [CrossRef]
59. Tüzel, Y.; Ekinci, K.; Öztekin, G.B.; Erdal, I.; Varol, N.; Merken, Ö. Utilization of olive oil processing waste composts in organic tomato seedling production. *Agronomy* **2020**, *10*, 797. [CrossRef]
60. Papafotiou, M.; Kargas, G.; Lytra, I. Olive-mill waste compost as a growth medium component for foliage potted plants. *HortScience* **2005**, *40*, 1746–1750. [CrossRef]
61. Ouzounidou, G.; Asfi, M.; Sotirakis, N.; Papadopoulou, P.; Gaitis, F. Olive mill wastewater triggered changes in physiology and nutritional quality of tomato (*Lycopersicon esculentum* Mill.) depending on growth substrate. *J. Hazard. Mater.* **2008**, *158*, 523–530. [CrossRef]
62. Kiarostami, K.; Mohseni, R.; Saboora, A. Biochemical changes of *Rosmarinus officinalis* under salt stress. *J. Stress Physiol. Biochem.* **2010**, *6*, 114–122.
63. Carmona, E.; Moreno, M.T.; Avilés, M.; Ordovas, J. Composting of wine industry wastes and their use as a substrate for growing soilless ornamental plants. *Span. J. Agric. Res.* **2012**, *10*, 482. [CrossRef]
64. Chrysargyris, A.; Maggini, R.; Incrocci, L.; Pardossi, A.; Tzortzakis, N. Copper tolerance and accumulation on *Pelargonium graveolens* l'hér. Grown in hydroponic culture. *Plants* **2021**, *10*, 1663. [CrossRef] [PubMed]

horticulturae

MDPI

Article

Hedypnois cretica L. and *Urospermum picroides* L. Plant Growth, Nutrient Status and Quality Characteristics under Salinity Stress

Alexios A. Alexopoulos [1,*], Anna Assimakopoulou [2], Panagiotis Panagopoulos [1], Maria Bakea [2], Nikolina Vidalis [3], Ioannis C. Karapanos [3], Youssef Rouphael [4,*] and Spyridon A. Petropoulos [5,*]

1 Laboratory of Agronomy, Department of Agriculture, University of the Peloponnese, Antikalamos, 24100 Kalamata, Greece
2 Laboratory of Plant Physiology, Department of Agriculture, University of the Peloponnese, Antikalamos, 24100 Kalamata, Greece
3 Laboratory of Vegetable Production, Department of Crop Science, Agricultural University of Athens, Iera Odos 75, 11855 Athens, Greece
4 Department of Agricultural Sciences, University of Naples Federico II, 80055 Portici, Italy
5 Department of Agriculture Crop Production and Rural Environment, University of Thessaly, Fytokou Street, 38446 Volos, Greece
* Correspondence: a.aleksopoulos@uop.gr (A.A.A.); youssef.rouphael@unina.it (Y.R.); spetropoulos@uth.gr (S.A.P.)

Abstract: The cultivation of tolerant species with low-quality irrigation water is one of the main strategies to address the lack of availability of irrigation water. In this work, the effect of salinity on plant growth, nutritional composition, and quality features of *Hedypnois cretica* and *Urospermum picroides* was tested. Fresh yield of leaves and roots of both species were severely decreased under high salinity, while the mineral profile of leaves and roots also showed a decrease in most minerals. The recorded values of K/Na and Ca/Na ratios indicate that *H. cretica* has a higher susceptibility to salt stress due to a higher decrease in the values of the respective ratios compared to *U. picroides*. Leaf pigments and total phenolic compounds content were not significantly affected by salinity, while total soluble solids and titratable acidity increased under increasing salinity in most cases. Moreover, free proline content increased with increasing salinity, while the opposite trend was recorded for nitrates' content. In conclusion, our findings indicate that tailor-made nutrient solutions could allow the use of irrigation water of a low quality for the commercial cultivation of *H. cretica* and *U. picroides*, thus allowing their domestication and integration in cropping systems where the cultivation of conventional crops is compromised.

Keywords: medicinal plants; total phenols; nutrient content; soilless cultivation; wild edible plants; scaly hawkbit; prickly goldenfleece

Citation: Alexopoulos, A.A.; Assimakopoulou, A.; Panagopoulos, P.; Bakea, M.; Vidalis, N.; Karapanos, I.C.; Rouphael, Y.; Petropoulos, S.A. *Hedypnois cretica* L. and *Urospermum picroides* L. Plant Growth, Nutrient Status and Quality Characteristics under Salinity Stress. *Horticulturae* **2023**, *9*, 65. https://doi.org/10.3390/horticulturae9010065

Academic Editor: Lijun Wang

Received: 12 December 2022
Revised: 24 December 2022
Accepted: 3 January 2023
Published: 5 January 2023

1. Introduction

The world population is constantly increasing and is predicted to reach 9.2 billion by 2050 [1]. Moreover, the percentage of people that face hunger or undernutrition globally increased from 8.4% in 2019 to 9.9% in 2020 mainly due to the COVID-19 pandemic, thus challenging the Zero Hunger target of 2030 [2]. At the same time, about 20% of agricultural land is affected by salinity [3], while it is predicted that the current soil salinization trends will result in 50% of arable land being salt-affected by 2050 [4], mainly because of poor agricultural practices and the impacts of climate change [5]. Therefore, food security is put under continuous threat directly or indirectly by anthropogenic activities.

To meet the increasing demands in food availability, it is imperative to exploit saline soils or even to reclaim arable lands that are not currently cultivated due to secondary salinity [6]. For this purpose, the adoption of proper agronomical practices and novel methods in mitigating abiotic stresses (e.g., use of biostimulants, microbial inoculations, etc.) could alleviate the negative effects of salinity on salt-sensitive crops [7]. However, this goal can be mainly achieved by breeding new cultivars of higher resistance to soil salinity or poor water quality, and by introducing new salt-resilient crops by domesticating wild, edible plants and halophytes, with proven resistance to salinity as well as to other abiotic stressors (e.g., water shortages, extreme temperatures, etc.) related to the increasing scarcity of irrigation water and climate change impacts [8].

In this context, many studies have been recently carried out to determine the level of salt-resistance in a number of wild, edible and medicinal plants that are traditionally consumed by gathering from the wild and could potentially serve as innovative/complementary crops, including *Cichorium spinosum* L. [9], *Diplotaxis tenuifolia* (L.) DC. [10], *Taraxacum officinale*, and *Reichardia picroides* (L.) Roth [11]. In addition, halophytes, such as *Crithmum maritimum* L. [12], *Chenopodium album* L. [13], *Plantago coronopus* L. [14], *Salsola komarovii* Iljin, *Sanguisorba minor* Scop. [15], and *Portulaca oleracea* L. [16], have been proposed as new horticultural or medicinal crops, due to their ability to adapt to harsh conditions such as irrigation with saline or sea-water, providing a solution to the continuous decrease in irrigation water availability [14]. Apart from being a valuable source of genotypes and genes that could be exploited in breeding programs towards integrating salt-resistance to crops, wild, edible and medicinal plants may also play an important role in daily diet and human health due to their richness in beneficial bioactive compounds [17,18]. In addition, saline conditions as well as other abiotic stresses may act as eustressors by inducing the biosynthesis of health-promoting compounds in a number of fresh vegetables, including wild, edible species [19].

Hedypnois cretica (L.) Dum. Cours (syn. *Hedypnois rhagadioloides* subsp. *cretica*) (Cretanweed or scaly hawkbit) and *Urospermum picroides* (L.) Scop. Ex F.W. Schmidt (syn. *Tragopogon picroides* L.) (prickly goldenfleece or prickly cupped goat's beard) are two wild, edible plants belonging to the Asteraceae family that are commonly consumed by local communities of the Mediterranean basin in raw, boiled, or cooked form and present a great potential for commercial cultivation as vegetables or medicinal species rich in health-promoting bioactive compounds [20]. However, being considered as an underutilized species, there is scarcity or even absence of information regarding their cultivation practices and their response to salinity in particular, although dunes and coastal areas are reported to be among the natural habitats of *H. cretica* [21], while *U. picroides* is considered to be NaCl-resistant up to 8 dS m^{-1} [22]. This knowledge is essential for the commercial exploitation of both species, especially in problematic soils, since the salinity thresholds have to be defined before introducing them as innovative crops. For instance, Alexopoulos et al. [11] recorded reduced growth in *Taraxacum officinale* (L.) Weber ex F.H. Wigg. plants cultivated in a floating system using NaCl-supplemented nutrient solution (EC 6 dS m^{-1}), whereas *R. picroides* plants were unaffected, probably due to its higher effectiveness in water uptake, accumulation of osmolytes (e.g., proline), and K/Na selectivity under salt-stress conditions.

Taking into account the lack of scientific data related to the response of these valuable, wild species to adverse environments, we studied the effect of NaCl-induced salinity at 6 and 10 dS m^{-1} on the growth, accumulation of nutrients, and quality/dietary traits of *H. cretica* and *U. picroides* plants, cultivated in a floating system aiming at their domestication and commercial exploitation in existing farming systems.

2. Materials and Methods

2.1. Plant Material, Experimental Treatments and Growth Conditions

The trial was carried out at the University of the Peloponnese (Kalamata, Messinia, Southern Greece, 37°3′22″ N, 22°1′43″ E) from November 2018 to March 2019, according to the methodology previously reported by Alexopoulos et al. [11]. In particular, seeds of Cretanweed (*Hedypnois cretica* (L.) Dum. Cours) and prickly goldenfleece (*Urospermum picroides* (L.) Scop. Ex F.W. Schmidt) were sown in November 2018 in germination containers (19 cm × 13 cm × 5 cm) containing white peat (pH 5.5–6.5, base substrate with no fertilizers added; Klasmann-Deilmann GmbH, Geeste, Germany). Then, the containers were put in a walk-in growth chamber (temperature = 20 °C, photoperiod 16 h, light intensity of fluorescent lamps = 55 μmol m^{-2} s^{-1}). When seedlings formed 3–4 true leaves (61 days after sowing; DAS), they were transferred to polystyrene seedling trays (cell dimension 5 cm × 5 cm × 5 cm) containing the same growing medium in a density of 44.44 plants m^{-2} (distances between plants 15 cm × 15 cm). The trays were transferred in an unheated glasshouse and put into containers filled with nutrient solution up to a height of 20 cm. The nutrient solution was continuously supplied with oxygen using air pumps. The composition of the nutrient solution was already described by Alexopoulos et al. [23]. The tested treatments included three nutrient solutions with different salinity levels, namely EC2, EC6, and EC10, prepared according to the procedure reported by Alexopoulos et al. [11,23]. During the experimental period, pH values were recorded daily and maintained and levels between 5.8 and 6.2, while EC values ranged from 2.0 to 2.2 dS m^{-1} (EC2), 6.0 to 6.2 dS m^{-1} (EC6), and 10.0 to 10.2 dS m^{-1} (EC10). Growing conditions were reported in the study of Alexopoulos et al. [11].

Both species were harvested 56 days after transplanting (DAT), when their rosette was comparable to the harvesting stage when collected in the wild (e.g., plants reached an adequate size being young and tender; Figure 1). The experiment for each species followed the completely randomized design, while four replications with ten plants each were implemented for each species (120 plants in total for each species).

Figure 1. Plants of *Urospermum picroides* (**A**) and *Hedypnois cretica* (**B**) grown at 2 dS m^{-1}.

2.2. *Plant Growth Parameters Measurement*

The number of leaves, the diameter of rosettes, length, and width of the largest leaf were recorded in both plant species at the harvesting date (56 DAT). Measurements were taken in all the plants of each replication (40 plants for each treatment) and after harvest, the plants of each replicate were sampled and grouped as one batch sample, i.e., four samples (replications) per treatment. The assessed parameters included the number of non-marketable leaves (yellowed, partially dried, or injured), the aboveground plant-parts' fresh weight (FW), root FW, and marketable leaf FW. After the assessment of the abovementioned parameters, the marketable leaves of each treatment were put together in a batch sample which was further split into two subsamples for further chemical analyses. One of the subsamples was used for the evaluation of dry matter content (% DMC) and the mineral profile determination of leaves, while the other subsample was kept at −80 °C for the assessment of chlorophyll, total soluble solids content (TSSC), total phenolic compounds (TPC), and proline content in the leaves. Similarly, the roots of each treatment were grouped as one sample, i.e., four samples (replications) per treatment, which was used for the determination of % DMC and of the macro-, and micro-elements' content. For the evaluation of % DMC, plant leaves and roots were air-dried at 72 °C to constant weight.

2.3. Determination of Mineral Profile in Leaves and Roots and Nitrate Concentration in Leaves

For the determination of total N, P, K, Ca, Mg, Na, Cl (expressed in % of plant tissue dry weight), and total Fe, Mn, Zn, Cu, and B (expressed in g kg^{-1} of plant tissue dry weight) contents in leaves and roots, the methodology previously reported by Alexopoulos et al. was followed [23]. The concentration of nitrates ($N\text{-}NO_3^-$) in leaves was determined according to the method reported by Cataldo et al. [24] and expressed in mg kg^{-1} of leaf FW.

Salinity tolerance index (STI) was calculated as the percent ratio of plant dry weight under salt stress and non-salt stress condition (relative plant dry-weight) using the following formula: Tolerance index (TI) (%) = (Treatment dry weight/Control dry weight) × 100 [25].

Regarding the nutrient accumulation, the nutrient contents per organ were calculated from the nutrient concentrations in the plant tissue (g Kg^{-1}) and the relevant dry matter (g per plant) and expressed as mg per plant.

2.4. Chemical Analyses in Leaves

Total soluble solids' content of leaves was evaluated with a portable refractometer (model HR32B; Schmidt + Haensch GmbH & Co., Berlin, Germany), as previously described by Alexopoulos et al. [11], while titratable acidity was determined by titration with NaOH [11].

Chlorophyll content of leaves was measured based on the readings of SPAD-502 Chlorophyll Meter (Konica-Minolta Co., Ltd., Tokyo, Japan) [23] and by quantification of chlorophylls in acetone extracts of leaves [26]. Carotenoids + xanthophylls were measured according to the method of Lichtenthaler and Buschmann [27].

Total phenolic compounds were quantified based on the Folin–Ciocalteu method with slight modifications [26], while free proline was assessed using the acid–ninhydrin method [28].

2.5. Statistical Analysis

The results for each species were analyzed with a one-way analysis of variance (ANOVA), while means comparison was performed with the Least Significant Difference (LSD) test at $p \leq 0.05$. The relationships between growth parameters and minerals contents and quality parameters were evaluated using Pearson's correlation test. All statistical tests were performed with the StatGraphics Centurion-XVI statistical package (StatPoint Technologies Inc., Warrenton, VA, USA).

3. Results and Discussion

3.1. Plant Growth Parameters

The leaf number of *H. cretica* was significantly decreased by increasing the salinity level (Table 1), whereas in the case of *U. picroides* the number of leaves was significantly higher in the EC2 than the EC10 treatment (Table 2). In particular, the percent leaf number decrease in *H. cretica* under EC6 treatment compared to the control (EC2) was similar to the decrease in *U. picroides* under EC10 treatment compared to the control (29.8% and 25.4%, respectively). The rosette diameter was severely affected by increasing salinity, especially in the case of *H. cretica* plants where high salinity (EC10) reduced the diameter of plants by 66.7%, whereas *U. picroides* was less affected showing a reduction of 35%. Moreover, the decrease in the number of leaves coincided with a significant increase in the number of non-marketable yield due to the loss of visual quality in both species (two-fold and three-fold increase in the case of EC10 compared to EC2 treatment, for *H. cretica* and *U. picroides*, respectively); however, the portion of unmarketable leaves in the total leaf number under high salinity was similar for both species (15% and 16% for *H. cretica* and *U. picroides*, respectively). Apart from the effects on the formation of new leaves, high salinity (EC10) significantly reduced the dimension of leaves, especially in *H. cretica* plants where the leaf length and width were reduced by 65.9% and 59.2%, respectively. On the other hand, no effects were recorded in the SPAD index for both species. Regarding yield-related parameters, total fresh weight was significantly reduced with increasing salinity for both species due to the formation of fewer and smaller leaves, while *H. cretica* seemed to be more susceptible to salinity stress compared to *U. picroides* (81.5% vs. 43.6% yield reduction under EC10 treatment, respectively). When allocating total plant weight in the aerial part and roots, a severe effect of high salinity level was also recorded, while the root/aerial part ratio revealed that the reduction in total plant weight was mostly due to the reduction in the weight of the aerial part, especially in the case of *H. cretica* plants. Finally, the percentage of leaf dry matter of *H. cretica* plants increased from EC2 to EC6 and from EC2 to EC10 by 16% and 46%, respectively, whereas that of *U. picroides* was not significantly impacted by the tested salinity levels (Tables 1 and 2).

Table 1. Growth parameters of *Hedypnois cretica* plants grown under different electrical conductivity (EC) levels (2, 6, and 10 dS m^{-1}).

EC (dS m^{-1})	Leaf Number Plant^{-1}	Rosette Diameter (cm)	Non-Marketable Leaf Number Plant^{-1}	SPAD Index	Maximum Leaf Length (cm)	Maximum Leaf Width (cm)
2	113.2 a	57.96 a	2.8 b	40.83 a	29.04 a	6.29 a
6	80.5 b	30.45 b	2.2 b	45.94 a	15.64 b	3.95 b
10	36.6 c	19.27 c	5.6 a	41.49 a	9.89 c	2.56 c
	Total Plant FW (g)	**Aerial Part FW (g)**	**Root FW (g)**	**Root/Aerial Part Ratio**	**Leaves FW (g) Plant^{-1}**	**Leaf DMC (%)**
2	132.47 a	105.92 a	26.55 a	0.25 b	94.83 a	7.26 c
6	51.54 b	33.23 b	18.32 b	0.55 a	30.70 b	8.42 b
10	24.49 c	13.66 c	10.83 c	0.82 a	10.70 c	10.62 a

Means within the same column followed by the same letter do not differ significantly based on Least Significant Difference (LSD) test at $p < 0.05$.

Table 2. Growth parameters of *Urospermum picroides* plants grown under different electrical conductivity (EC) levels (2, 6, and 10 dS m^{-1}).

EC (dS m^{-1})	Leaf Number Plant^{-1}	Rosette Diameter (cm)	Non-Marketable Leaf Number Plant^{-1}	SPAD Index	Maximum Leaf Length (cm)	Maximum Leaf Width (cm)
2	25.2 a	39.46 a	1.00 b	41.32 b	20.49 a	6.07 a
6	23.0 ab	30.45 b	1.68 b	48.82 a	15.74 b	4.91 b
10	18.8 b	25.64 b	3.08 a	48.61 a	13.33 b	4.40 b
	Total Plant FW (g)	**Aerial Part FW (g)**	**Root FW (g)**	**Root/Aerial Part Ratio**	**Leaves FW (g) Plant^{-1}**	**Leaf DMC (%)**
2	55.70 a	41.24 a	14.46 a	0.35 a	36.62 a	8.28 a
6	42.70 b	29.95 b	12.74 ab	0.43 a	25.51 b	8.69 a
10	31.43 c	22.60 c	8.83 b	0.39 a	19.92 b	8.64 a

Means within the same column followed by the same letter do not differ significantly based on Least Significant Difference (LSD) test at $p < 0.05$.

Similarly to our study, Uddin et al. [29] suggested a reduction in the number of leaves of three *Brassica* species subjected to salt stress, while Klados and Tzortzakis also reported a significant decrease in the number of leaves of *Cichorium spinosum* plant grown under saline conditions. Moreover, taking into consideration the plant leaf FW decrease in combination with the maximum leaf length and width decrease under salt stress, we could assume that salinity affected negatively the leaf expansion, thus limiting the photosynthetically active surface area and the biosynthesis of structural compounds that allowed for the formation of new leaves and the development of the existing ones [30]. The severe effects of salinity on fresh biomass yield were also reported in the case of *Cichorium intybus* L. where the aerial part and roots showed similar decreasing trends. This finding coincides with the results of our study regarding *U. picroides* and not with those of *H. cretica* highlighting the significant effect of genotypes on salinity stress [31].

The higher tolerance of *U. picroides* to salinity stress could be attributed to better water uptake, as indicated by the non-significant impact of the tested salinity levels on leaf dry matter content and consequently to water content of leaves. It has been well-documented that the presence of high amounts of NaCl in the nutrient solution may result in hyperosmotic conditions which hamper plant absorption of water and nutrients [32]. To the authors' knowledge, there are no available reports on the tolerance of the tested species to salinity stress. However, in other wild species such as *Taraxacum erythropodium* Kitag., plants grown under high soil salt content (>0.7%) also showed reduced growth [33], while in the case of *Reichardia picroides* the plants were resistant to high salinity levels based on the recorded growth parameters' values [11,22].

3.2. Leaf and Root Mineral Profile

Leaf N concentration of *H. cretica* was not significantly affected by increasing EC levels in comparison to the control, whereas in regard to *U. picroides* the N content was significantly lower for the EC6 treatment compared either to the control (EC2) or the high salinity level (EC10) (Table 3). This variable response of the tested species could be attributed to the severe decrease in the leaf fresh biomass in the case of *H. cretica* compared to *U. picroides* under the same salt conditions. Moreover, the variability of the salt stress impact on N metabolism could be also attributed to genotypic differences between the tested plant species which suggest the variability in the effectiveness of protective mechanisms against abiotic stressors [11,34].

Table 3. Leaf nutrient concentrations of *Hedypnois cretica* (3a) and *Urospermum picroides* (3b) plants grown under different nutrient solution EC levels (2, 6, and 10 dS m^{-1}).

EC	N	P	K	Ca	Mg	Na	Cl	Fe	Mn	Zn	Cu	B	K/Na	Ca/Na
(dS m^{-1})			(mg g^{-1} Leaf DW)								(mg kg^{-1} Leaf DW)			
							Hedypnois cretica							
2	44.6 a	8.1 b	67.0 a	11.0 a	2.2 a	3.80 c	5.0 c	72.5 a	108.1 a	99.3 a	8.7 a	153.1 a	18.4 a	2.99 a
6	44.7 a	13.4 a	35.0 b	7.1 b	2.1 a	3.84 b	15.0 b	76.9 a	58.3 b	112.3 a	9.8 a	51.7 b	0.9 b	0.19 b
10	44.3 a	13.3 a	30.0 b	5.9 b	2.1 a	49.7 a	24.0 a	74.2 a	68.2 b	114.8 a	9.2 a	48.8 b	0.6 b	0.12 b
							Urospermum picroides							
2	46.2 a	8.8 b	62.0 a	8.7 a	2.1 a	9.6 c	6.0 c	69.2 b	193.3 a	117.3 a	7.0 a	142.7 b	6.4 a	0.90 a
6	43.2 b	9.4 b	41.0 b	5.0 b	1.6 b	34.4 b	14.0 b	76.3 ab	71.7 b	88.5 b	6.6 a	152.0 ab	1.2 b	0.15 b
10	48.7 a	12.2 a	30.0 c	4.3 b	1.7 b	46.8 a	20.0 a	78.3 a	85.8 b	95.7 b	7.7 a	156.5 a	0.6 c	0.09 b

Means within the same column (for each plant species separately) followed by the same letter do not differ significantly according to the Least Significant Difference (LSD) test at $p < 0.05$.

Leaf P concentration of both species was significantly increased in *H. cretica* plants for both salinity levels (e.g., EC6 and EC10) in comparison to the control treatment (EC2), whereas in the case of *U. picroides*, the P content increased only under the high salinity as compared to EC2 treatment (Table 3). Moreover, P content in roots was significantly higher in plants subjected to EC6 and EC10 levels in comparison to the control treatment for both tested species (Table 4). Similar results were reported for the edible, wild species *Taraxacum officinale* and *Reichardia picroides* by Alexopoulos et al. [11] who used similar growing conditions (nutrient solution composition, light conditions, salinity levels) to our study. In contrast, Assimakopoulou et al. [35] recorded a decrease in the P content of tomato leaves, while Grattan and Grieve [36] reported that salinity either improved or had no effect on P uptake in horticultural crops. These differences could be attributed to the different species tested, since the importance of genotype to response of species to salinity stress is well-established [36]. Therefore, it could be suggested that the response of plants regarding the uptake of P under saline conditions is genotype-dependent, while growing conditions may also have an impact on plants' nutrient profile.

Table 4. Root nutrient concentrations at harvest date of *Hedypnois cretica* and *Urospermum picroides* plants grown under different nutrient solution EC levels (2, 6, and 10 dS m^{-1}).

EC	P	K	Ca	Mg	Na	Fe	Mn	Zn	Cu	B
(dS m^{-1})		($mg\ g^{-1}$ Leaf DW)					($mg\ kg^{-1}$ Root DW)			
					Hedypnois cretica					
2	13.7 b	72.0 a	4.2 b	1.4 a	1.6 c	267.4 a	53.1 a	92.4 a	41.3 a	28.0 a
6	16.3 a	52.0 b	5.4 ab	1.7 a	21.6 b	162.5 b	15.2 c	73.1 b	30.9 b	24.7 a
10	15.6 a	40.0 c	5.9 a	1.7 a	29.0 a	144.1 b	22.5 b	92.5 a	20.8 c	22.6 a
					Urospermum picroides					
2	12.9 b	63.0 a	4.1 a	1.9 a	1.3 c	119.8 b	123.1 a	215.4 a	12.8 b	31.9 a
6	15.2 a	64.0 a	3.8 a	1.5 ab	11.6 b	149.7 a	70.3 b	117.5 b	16.1 a	26.5 b
10	16.3 a	48.0 b	2.6 a	1.2 b	21.4 a	79.0 c	26.7 c	104.6 b	12.4 b	25.2 b

Means within the same column (for each plant species separately) followed by the same letter do not differ significantly according to the Least Significant Difference (LSD) test at $p < 0.05$.

For both species, leaf K and Ca content decreased significantly under the two salinity levels in comparison to the control treatment without significant differences being recorded between them (except for the case of *U. picroides* where leaf K content in EC10 treatment was significantly different from EC6 treatment) (Table 3). Leaf Mg content was not significantly affected by increasing salinity in *H. cretica*, whereas a significant reduction was recorded in both salinity levels in comparison to the control treatment (Table 3). In regard to roots, the Mg content was not significantly affected in *H. cretica* plants, while a significant decrease was recorded only at the highest salinity level tested (EC10) compared to the control treatment (Table 4). This decrease in leaf K and Ca content under salt stress could be associated with the known antagonistic effects between Na^+, K^+, and Ca^{2+}, since they probably share the same transport systems either at the root surface or into the xylem [37], while the high content of Na^+ inhibits the uptake process of K^+ and Ca^{2+} [38] with severe effects on plant growth and yield in various crops [39]. Alexopoulos et al. [11] also reported a decreasing trend in K and Ca content of leaves in *Taraxacum officinale* and *Reichardia picroides*, a finding which confirms our results, especially when considering the similarities in the growing and experimental conditions between these two studies. In contrast, other researchers have reported non-significant effects of salinity on leaf K and Mg contents in pepper [40], while Assimakopoulou et al. [35] found increased leaf Ca, Mg, and K content in tomato plants. These contradictory reports could be due to the variable effect of salinity stress on plants, depending on several factors such as growing conditions, growth medium, stress duration, plant genotype, as well as to vegetable type (e.g., leafy or fruit vegetable) [41].

In the case of roots, the root K content was gradually decreased with increasing salinity in *H. cretica* plants, while in *U. picroides* the K content decreased only at the high salinity level (Table 4). According to Zhang et al. [42], the K content is pivotal for the homeostasis in salt tolerance mechanisms, which indicates more effective stress protective mechanisms in *U. picroides* than *H. cretica* plants. Moreover, our results indicate that the decrease in K content was more profound in leaves than in roots in both species, a finding which coincides with the reports of Pérez-Alfocea et al. [43] in tomato plants or the results of Alexopoulos et al. [11] in *Taraxacum officinale* and *Reichardia picroides* plants.

Leaf Na and Cl content, as well as root Na content recorded a gradual increase with increasing salinity in both species, due to the increasing availability of both minerals in nutrient solution in saline conditions (Tables 3 and 4). Moreover, considering that *U. picroides* plants had lower Na content in leaves and roots compared to *H. cretica* plants, it could be suggested that the observed tolerance of the former species could be due to Na excretion mechanisms [44,45].

Moreover, both the leaf K/Na and Ca/Na ratios reflected the relevant salt tolerance of the studied species in the present work, as in *H. cretica* a higher decrease in K/Na and Ca/Na ratios in the cytosol was recorded compared to *U. picroides*, suggesting a higher susceptibility to salt stress in the former species. Similar results were suggested by other researchers who also suggested the ratios of K/Na and Ca/Na as indicative indices for tolerance to salinity stress in various crops [46]. According to Mousavi et al. [47], the K/Na reflects the effectiveness of the K–Na exchange in plasmalemma which is responsible for salinity tolerance through the removal of Na from the cytoplast and its decreased translocation through the xylem.

Regarding micronutrients, leaf Fe, Zn, and Cu content was not significantly impacted by salinity levels in *H. cretica* plants, whereas Mn and B were negatively affected by saline conditions (Table 3). On the other hand, EC6 and EC10 treatments had no significant effect on leaf Cu content of *U. picroides*, while Mn and Zn content significantly decreased under salinity treatments. Finally, the content of Fe and B in leaves showed an increase over the control treatment when plants were grown at high salinity levels (EC10). As shown in Table 4, the micronutrients' content in *H. cretica* roots was significantly lower compared to the control plants under the salt treatments EC6 and EC10; however, Zn content decreased only at EC6 and B which was not significantly affected by the tested treatments. Similarly, a negative impact of salinity on micronutrients content was recorded for *U. picroides*, regardless of the salinity level, except for the case of Fe and Cu where the EC6 level resulted in a significant increase compared to the rest of the treatments. This varied response to salinity levels has also been confirmed in the literature studies where inconsistent results regarding the micronutrients content under saline conditions have been suggested in various crops, indicating the importance of genotype as well as the experimental treatments (salinity level, stress duration, stress initiation etc.) and growing conditions (cropping season, air temperature, light conditions) [48].

3.3. Salinity Tolerance Index

In order to assess the tolerance of the tested species, the salinity tolerance index was determined based on the recorded values of dry weight of plants in the control (EC2) and the studied salinity levels (Figure 2). Comparing the tolerance index of the two plants, *U. picroides* and the control treatment, it could be suggested that *U. picroides* was more tolerant than *H. cretica* in either the EC6 or EC10 treatments. According to the literature, there is a strong correlation between Na content in leaves, with tolerant genotypes showing a lower accumulation of Na than less tolerant ones [49]. This finding agrees with our study, since the increase in Na content in leaves and roots under high salinity levels (EC10) was lower in the case of *U. picroides* than *H. cretica* (increased by 3.85 times vs. 12.08 times in the case of leaves for *U. picroides* and *H. cretica*, respectively; and 15.46 times vs. 17.12 times in the case of roots for *U. picroides* and *H. cretica*, respectively). Moreover, Tao et al. [50] reported a wide variability between different wheat genotypes in values of the salinity tolerance index while Anshori et al. [51] highlighted the importance of using various salinity tolerance indices for the selection of elite rice genotypes in breeding programs.

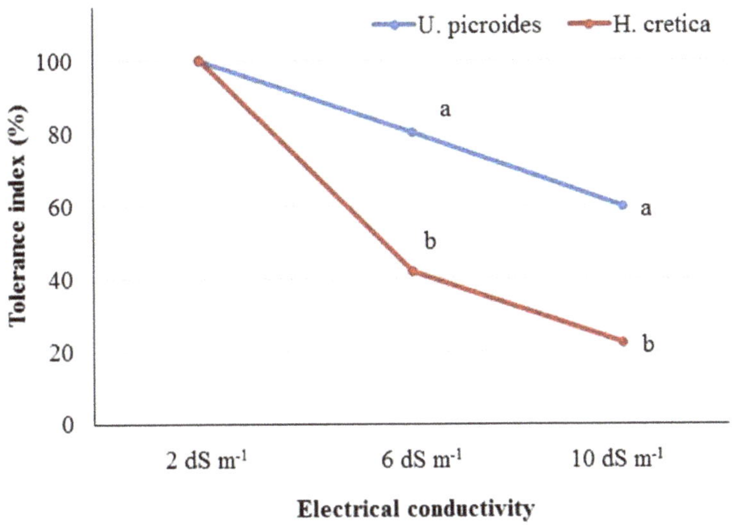

Figure 2. Salinity tolerance index of *Urospermus picroides* and *Hedypnois cretica* plants grown under different EC levels (2, 6, and 10 dS m^{-1}) in nutrient solution. Different letters indicate statistically significant differences in the salinity tolerance index of the plant species (in each EC level separately) according to the *t*-test at $p < 0.05$.

3.4. Quality Traits (Leaf Pigments, Total Soluble Solids Content, Titratable Acidity, and Total Phenolic Compounds Content)

In both species, salinity had no effect on the content of leaves in chlorophylls (chl a, chl b, and total chlorophylls) as well as in carotenoids, xanthophylls, and total phenols (Table 5). The lack of effect of salinity levels on chlorophyll content is in accordance with our results for SPAD index values presented in Tables 1 and 2. According to the literature, moderate to high salinity levels similarly had no effect on the content of pigments in the leaves of *Taraxacum officinale* and *Reichardia picroides* (up to 10 dS m^{-1}; [11]), and leaf chlorophyll content of *Cichorium spinosum* (up to 40 mM NaCl; [52]). In contrast, a gradual addition of NaCl/$CaCl_2$ up to 30/15 mM (EC 8.4 dS m^{-1}) in lettuce genotypes [53] and up to 20/10 mM (EC 6.5 dS m^{-1}) in spinach [54] resulted in increased chlorophyll content (a, b, and total) and chlorophyll a/b ratio. However, although both lettuce and spinach are considered moderately sensitive to salinity [31], the evaluation of 178 cultivars and germplasm accessions of lettuce resulted in either no change or in an increase in the leaf SPAD units, thus indicating a genotype-dependent response to salinity stress. In the case of *Taraxacum erythropodium*, soil salt contents above 0.4% had a negative effect on chlorophyll content [33]. Therefore, it could be suggested that depending on the genotype, the application of mild or relatively high salinity stress do not negatively affect the content of leafy vegetables in pigments and thus do not impair their coloration, an organoleptic trait of particular importance for their marketability and preference by consumers [55]. Moreover, apart from the involvement of leaf pigments in physiological functions (i.e., photosynthesis), they are also beneficial for human health, as they mitigate oxidative stress and help in the prevention of chronic diseases [56]. Therefore, the lack of effect of salinity on leaf pigments indicate that visual and nutritional quality of the edible product of both species is not compromised by saline conditions.

Table 5. Chlorophyll (a, b, and total; mg/100 g FW); carotenoids + xanthophylls (mg/100 g FW); total phenolics (mg GAE/100 g FW); total soluble solids content (TSSC; °Brix); titratable acidity (TA; g malic acid/100 g FW); leaf nitrate content (mg/kg fw) and free proline content (μmole/g FW) of *Hedypnois cretica* and *Urospermum picroides* plants grown under different nutrient solution EC (2, 6, and 10 dS m^{-1}) levels.

EC	Chlorophyll a	Chlorophyll b	Total Chlorophyll	Carotenoids+ Xanthophylls	Total Phenolic Compounds	TSSC	TA	Nitrate Content	Free Proline
(dS m^{-1})		(mg/100 g FW)			(mg GAE /100 g FW)	°Brix	g Malic Acid/100 g FW	(mg/kg FW)	(μmole/g FW)
					Hedypnois cretica				
2	49.22 a	24.15 a	73.37 a	6.90 a	52.66 a	3.75 b	0.043 c	4317.5 a	0.006 b
6	51.58 a	24.90 a	76.48 a	7.35 a	58.90 a	4.23 ab	0.115 b	2323.2 b	0.028 b
10	50.10 a	24.70 a	74.80 a	6.42 a	50.35 a	5.03 a	0.184 a	1311.3 c	1.444 a
					Urospermum picroides				
2	55.48 a	25.89 a	81.37 a	8.29 a	62.60 a	3.85 b	0.178 a	2534.2 a	0.046 c
6	52.22 a	24.43 a	76.65 a	8.36 a	73.11 a	4.28 ab	0.171 a	1057.3 b	0.606 b
10	57.05 a	26.28 a	83.33 a	8.51 a	70.82 a	4.45 a	0.204 a	993.6 b	2.339 a

Means within the same column (for each plant species separately) followed by the same letter do not differ significantly according to the LSD test at $p < 0.05$.

Leaf content in total phenolic compounds (TPC) was not significantly impacted by the tested salinity treatments in both species (Table 5). Although it is reported that salinity enhances the biosynthesis of phenolic compounds in plants due to the induction of secondary metabolism [57], several other factors could be involved in this response such as the salt dose, the initiation of salt treatment in relation to plant developmental stage, the duration of the treatment, and the source of salt stress among others [58]. Thus, in agreement with our results, no change in TPC was recorded in romaine-type lettuce under a long-term mild NaCl stress (5 mM NaCl; [59]), in green and red-leafed baby lettuce grown at 10 mM NaCl [60], in spinach after a combined NaCl/CaCl2 salt-treatment at 6.5 dS m^{-1} [54], or in spiny chicory (*Cichorium spinosum*) after NaCl treatment up to 8 dS m^{-1} [9]. On the contrary, lettuce plants subjected at 50 mmol L^{-1} NaCl had higher TPC and phenolic acids' content, whereas a more severe stress (150 mmol L^{-1} NaCl) favored the accumulation of flavonoids [61]. Similarly, Klados and Tzortzakis [62] observed an increase in TPC at 120 mM NaCl in spiny chicory plants. However, although TPC and antioxidant compounds such as carotenoids were not affected under the studied conditions, the determination of other bioactive compounds could provide useful information regarding the potential of implementing saline irrigation water as a cost-effective means for increasing functional quality of wild, edible species, as already indicated on other occasions [9,63].

Leaf total soluble-solids' content (TSSC) increased in both species after the application of EC10 treatment in comparison to the control plants, whereas non-significant differences were recorded between EC6 and the control treatment or between EC6 and EC10 treatments (Table 5). In other wild, edible species such as *Taraxacum officinale* and *Reichardia picroides*, the TSSC values were unaffected by salinity levels up to 10 dS m^{-1} [11], whereas salt-treated *Cichorium spinosum* plants showed a reduced sugar accumulation [9]. In the case of lettuce plants, TSSC values either increased at mild salt-stress [64] or showed no significant changes at higher salinity levels [65], while Sakamoto et al. [66] reported a two-fold increase in TSSC values at NaCl-supplemented nutrient solution up to EC 12.6 dS m^{-1} in red lettuce leaves. Moreover, titratable acidity (TA) of *H. cretica* leaves gradually increased with increasing EC levels, whereas no significant differences were recorded in the case of *U. picroides*. The higher titratable acidity observed in *H. cretica* leaves under both salinity treatments could be due to concentration effect, as salinity (especially at 10 dS m^{-1}) was detrimental for plant growth, resulting in a significant increase in leaf DMC and consequently to lower water content [67].

3.5. Proline Content

The application of salinity treatments resulted in a gradual increase in the proline content compared to the control treatment only in the case of *U. picroides*, whereas the content in *H. cretica* leaves was increased only when the plants were subjected to high salinity (EC10) (Table 5). Given that proline is related to osmotic adjustment when plants are subjected to salinity stress [34], the higher salt tolerance of *U. picroides* indicated by the growth parameters (see results in Table 2) could be associated with the induction of proline biosynthesis, especially under the highest salinity levels tested (EC10). This finding agrees with the literature reports regarding *Brassica napus* L. [55], *Taraxacum erythropodium* [33], and strawberry [68] where the content of leaf proline showed an increase under increased salinity levels. Moreover, it seems that the protective mechanism of the species is triggered even at moderate salinity levels (EC6), whereas in the case of *H. cretica*, proline biosynthesis was significantly increased only when the plants were subjected to high salinity. Therefore, the lower fresh biomass yield recorded for *H. cretica* plants under saline conditions (see results in Table 1) could be associated with the inefficient protective mechanisms of the species and the biosynthesis of adequate amounts of proline. This difference in proline biosynthesis between the tested species could be attributed to the genetic manipulation of proline metabolism and the accumulation of the species [69].

3.6. Leaf Nitrate Concentration

The leaf nitrate concentration of the control plants in both plant species was found to be significantly higher than salinity treatments, while a gradual decrease with increasing salinity was recorded in *H. cretica* plants (Table 5). On the other hand, no significant differences in nitrate content were recorded between EC6 and EC10 in the case of *U. picroides*. *H. cretica* presented 1.7, 2.2, and 1.3 times higher leaf nitrate content under EC2, EC6, and EC10, respectively, as compared to the respective salinity levels in *U. picroides* leaves. Human exposure to nitrate is receiving significant attention from regulators and governmental officers due to the potential harmful effects on human health [70]. In particular, leafy vegetables and specific species such as spinach and lettuce are considered as potentially harmful due to nitrate accumulation [71].

The synthesis and accumulation of nitrates in leafy vegetables has been associated with several factors: (i) genotype, (ii), agronomic practices such as fertilizer rates, application form and timing of nitrogen, and (iii) growing conditions such as light intensity and quality, air temperature, and concentration of carbon dioxide [19,70]. Moreover, due to antagonism between nitrates and chloride, salinity elicited through a NaCl addition in nutrient solution has been suggested as an effective practice to reduce nitrate accumulation in leafy vegetables [19,70]. Therefore, replacing nitrate with chloride in the feeding solution in hydroponic cropping systems, could lead to the reduction in anti-nutrient nitrates in leafy vegetables [72]. For instance, Colla et al. [73] found that increasing the NaCl in the nutrient solution from 1 to 30 mM NaCl decreased the nitrogen content in leaves of various artichoke and cardoon cultivars grown in a floating system. Moreover, the reduction in nitrogen content at saline conditions was followed by a significant increase in antioxidant activity and polyphenols' profile. In addition, the decrease in the NO_3^-:Cl^- ratio in the feeding solution impaired translocation of N and NO_3^- in the shoots of cardoon plants cultivated in a floating system, but increased the total antioxidant activity, flavonoids and phenols' content in leaves [74]. The reduction in nitrate uptake could be attributed to a high chloride content, since the translocation of nitrates from roots to shoots might be reduced at the site of entrance into the xylem due the competition for the same anion channel [75].

4. Conclusions

Our results suggest a significant impact of electrical conductivity levels of nutrient solution on plant growth parameters and quality traits of *Hedypnois cretica* and *Urospermum picroides*. Although fresh biomass yield was severely affected in both species, *U. picroides* seems to be more resilient than *H. cretica*. This finding is also confirmed by the mineral profile of leaves and the values of K/Na and Ca/Na ratios, as well as the calculated stress tolerance index and the free proline content which all support the argument for *U. picroides* possessing a higher salinity stress tolerance. The evaluation of quality traits revealed that the tested salinity treatments had no significant effect on leaf pigments and total phenolic compounds content, whereas titratable acidity and total soluble solids content increased under salinity treatments indicating positive effects on the taste of the edible products. Moreover, it is worth mentioning that the content of nitrates was significantly reduced even under moderate salinity levels, which is also a beneficial effect on the quality of the edible product. In conclusion, our results indicate that the studied species showed a varied response to salinity treatments, with *U. picorides* being more tolerant to salt stress than *H. cretica*. Moreover, the application of tailor-made nutrient solutions may improve the quality traits of the edible portion of both species without compromising yield parameters, thus suggesting a cost-effective practice to increase the added value of both species and exploit them in commercial cropping systems.

Author Contributions: Conceptualization, A.A.A., I.C.K. and S.A.P.; methodology, P.P., M.B., A.A. and N.V.; formal analysis, P.P., M.B., A.A. and N.V.; investigation, P.P., M.B., A.A. and N.V.; resources, A.A.A., I.C.K. and S.A.P.; data curation, A.A.A. and I.C.K.; writing—original draft preparation, A.A.A. and S.A.P.; writing—review and editing, A.A.A., Y.R., I.C.K. and S.A.P.; visualization, A.A.A.; supervision, A.A.A., I.C.K. and S.A.P.; project administration, A.A.A. and S.A.P.; funding acquisition, S.A.P. All authors have read and agreed to the published version of the manuscript.

Funding: This research received no external funding.

Institutional Review Board Statement: Not applicable.

Informed Consent Statement: Not applicable.

Data Availability Statement: Data are available upon request.

Conflicts of Interest: The authors declare no conflict of interest.

References

1. Stephenson, J.; Newman, K.; Mayhew, S. Population dynamics and climate change: What are the links? *J. Public Health* **2010**, *32*, 150–156. [CrossRef] [PubMed]
2. FAO; IFAD; UNICEF; WFP; WHO. The State of Food Security and Nutrition in the World 2021. In *Transforming Food Systems for Food Security, Improved Nutrition and Affordable Healthy Diets for All*; FAO: Rome, Italy, 2021; p. 240.
3. FAOSTAT Production and Trade Statistics. Available online: http://www.fao.org/faostat/en/#data/QC/visualize (accessed on 24 December 2022).
4. Ashraf, M.; Foolad, M.R. Roles of glycine betaine and proline in improving plant abiotic stress resistance. *Environ. Exp. Bot.* **2007**, *59*, 206–216. [CrossRef]
5. Jamil, A.; Riaz, S.; Ashraf, M.; Foolad, M.R. Gene expression profiling of plants under salt stress. *CRC. Crit. Rev. Plant Sci.* **2011**, *30*, 435–458. [CrossRef]
6. Petropoulos, S.A.; Karkanis, A.; Martins, N.; Ferreira, I.C.F.R. Edible halophytes of the Mediterranean basin: Potential candidates for novel food products. *Trends Food Sci. Technol.* **2018**, *74*, 69–84. [CrossRef]
7. Rouphael, Y.; Petropoulos, S.A.; Cardarelli, M.; Colla, G. Salinity as eustressor for enhancing quality of vegetables. *Sci. Hortic.* **2018**, *234*, 361–369. [CrossRef]
8. Fita, A.; Rodríguez-Burruezo, A.; Boscaiu, M.; Prohens, J.; Vicente, O. Breeding and domesticating crops adapted to drought and salinity: A new paradigm for increasing food production. *Front. Plant Sci.* **2015**, *6*, 978. [CrossRef]
9. Petropoulos, S.; Levizou, E.; Ntatsi, G.; Fernandes, Â.; Petrotos, K.; Akoumianakis, K.; Barros, L.; Ferreira, I. Salinity effect on nutritional value, chemical composition and bioactive compounds content of *Cichorium spinosum* L. *Food Chem.* **2017**, *214*, 129–136. [CrossRef]
10. Bonasia, A.; Lazzizera, C.; Elia, A.; Conversa, G. Nutritional, biophysical and physiological characteristics of wild rocket genotypes as affected by soilless cultivation system, salinity level of nutrient solution and growing period. *Front. Plant Sci.* **2017**, *8*, 300. [CrossRef]
11. Alexopoulos, A.A.; Assimakopoulou, A.; Panagopoulos, P.; Bakea, M.; Vidalis, N.; Karapanos, I.C.; Petropoulos, S.A. Impact of salinity on the growth and chemical composition of two underutilized wild edible greens: *Taraxacum Officinale* and *Reichardia picroides*. *Horticulturae* **2021**, *7*, 160. [CrossRef]
12. Renna, M. Reviewing the prospects of sea fennel (*Crithmum maritimum* L.) as emerging vegetable crop. *Plants* **2018**, *7*, 92. [CrossRef]

13. Calone, R.; Bregaglio, S.; Sanoubar, R.; Noli, E.; Lambertini, C.; Barbanti, L. Physiological adaptation to water salinity in six wild halophytes suitable for Mediterranean agriculture. *Plants* **2021**, *10*, 309. [CrossRef]
14. Koyro, H.-W. Effect of salinity on growth, photosynthesis, water relations and solute composition of the potential cash crop halophyte *Plantago coronopus* (L.). *Environ. Exp. Bot.* **2006**, *56*, 136–146. [CrossRef]
15. Karkanis, A.C.; Fernandes, A.; Vaz, J.; Petropoulos, S.; Georgiou, E.; Ciric, A.; Sokovic, M.; Oludemi, T.; Barros, L.; Ferreira, I. Chemical composition and bioactive properties of *Sanguisorba minor* Scop. under Mediterranean growing conditions. *Food Funct.* **2019**, *10*, 1340–1351. [CrossRef]
16. Karkanis, A.C.; Petropoulos, S.A. Physiological and growth responses of several genotypes of common purslane (*Portulaca oleracea* L.) under Mediterranean semi-arid conditions. *Not. Bot. Horti Agrobot. Cluj-Napoca* **2017**, *45*, 569–575. [CrossRef]
17. Sánchez-Mata, M.C.; Loera, R.D.C.; Morales, P.; Fernández-Ruiz, V.; Cámara, M.; Marqués, C.D.; Pardo-de-Santayana, M.; Tardío, J. Wild vegetables of the Mediterranean area as valuable sources of bioactive compounds. *Genet. Resour. Crop Evol.* **2012**, *59*, 431–443. [CrossRef]
18. Chatzopoulou, E.; Carocho, M.; Di Gioia, F.; Petropoulos, S.A. The beneficial health effects of vegetables and wild edible greens: The case of the mediterranean diet and its sustainability. *Appl. Sci.* **2020**, *10*, 9144. [CrossRef]
19. Rouphael, Y.; Kyriacou, M.C.; Petropoulos, S.A.; De Pascale, S.; Colla, G. Improving vegetable quality in controlled environments. *Sci. Hortic.* **2018**, *234*, 275–289. [CrossRef]
20. Petropoulos, S.A.; Fernandes, Â.; Tzortzakis, N.; Sokovic, M.; Ciric, A.; Barros, L.; Ferreira, I.C.F.R. Bioactive compounds content and antimicrobial activities of wild edible Asteraceae species of the Mediterranean flora under commercial cultivation conditions. *Food Res. Int.* **2019**, *119*, 859–868. [CrossRef]
21. Badalonas, D.; Sykora, K.; Papastergiadou, E. Review of plant communities from Greek dunes and salt marshes, a preliminary summarizing list. *Ann. Di Bot.* **1995**, *53*, 107–118.
22. Salonikioti, A.; Petropoulos, S.; Antoniadis, V.; Levizou, E.; Alexopoulos, A. Wild Edible Species with Phytoremediation Properties. *Procedia Environ. Sci.* **2015**, *29*, 98–99. [CrossRef]
23. Alexopoulos, A.A.; Marandos, E.; Assimakopoulou, A.; Vidalis, N.; Petropoulos, S.A.; Karapanos, I.C. Effect of Nutrient Solution pH on the Growth, Yield and Quality of *Taraxacum officinale* and *Reichardia picroides* in a Floating Hydroponic System. *Agronomy* **2021**, *11*, 1118. [CrossRef]
24. Cataldo, D.A.; Maroon, M.; Schrader, L.E.; Youngs, V.L. Rapid colorimetric determination of nitrate in plant tissue by nitration of salicylic acid. *Commun. Soil Sci. Plant Anal.* **1975**, *6*, 71–80. [CrossRef]
25. Bálint, A.F.; Kovács, G.; Sutka, J. Copper tolerance of Aegilops, Triticum, Secale and triticale seedlings and copper and iron content in their shoots. *Acta Biol. Szeged.* **2002**, *46*, 77–78.
26. Karapanos, I.; Papandreou, A.; Skouloudi, M.; Makrogianni, D.; Fernández, J.A.; Rosa, E.; Ntatsi, G.; Bebeli, P.J.; Savvas, D. Cowpea fresh pods—A new legume for the market: Assessment of their quality and dietary characteristics of 37 cowpea accessions grown in southern Europe. *J. Sci. Food Agric.* **2017**, *97*, 4343–4352. [CrossRef] [PubMed]
27. Lichtenthaler, H.K.; Buschmann, C. Chlorophylls and Carotenoids: Measurement and Characterization by UV-VIS Spectroscopy. *Curr. Protoc. Food Anal. Chem.* **2001**, *1*, F4.3.1–F4.3.8. [CrossRef]
28. Bates, L.S. Rapid determination of free proline for water-stress studies. *Plant Soil* **1973**, *39*, 205–207. [CrossRef]
29. Uddin, M.N.; Tariqul Islam, M.; Karim, M.A. Salinity tolerance of three mustard/rapeseed cuitivars. *J. Bangladesh Agric. Univ.* **2005**, *3*, 203–208. [CrossRef]
30. Wang, Y.; Nii, N. Changes in chlorophyll, ribulose bisphosphate carboxylase-oxygenase, glycine betaine content, photosynthesis and transpiration in Amaranthus tricolor leaves during salt stress. *J. Hortic. Sci. Biotechnol.* **2000**, *75*, 623–627. [CrossRef]
31. Shannon, M.C.; Grieve, C.M. Tolerance of vegetable crops to salinity. *HortScience* **1999**, *78*, 5–38. [CrossRef]
32. Ryu, H.; Cho, Y.G. Plant hormones in salt stress tolerance. *J. Plant Biol.* **2015**, *58*, 147–155. [CrossRef]
33. Wu, Z.; Xue, Z.; Li, H.; Zhang, X.; Wang, X.; Lu, X. Cultivation of dandelion (*Taraxacum erythropodium*) on coastal saline land based on the control of salinity and fertilizer. *Folia Hortic.* **2019**, *31*, 277–284. [CrossRef]
34. Ashraf, M. Biotechnological approach of improving plant salt tolerance using antioxidants as markers. *Biotechnol. Adv.* **2009**, *27*, 84–93. [CrossRef]
35. Assimakopoulou, A.; Nifakos, K.; Salmas, I.; Kalogeropoulos, P. Growth, Ion Uptake, and Yield Responses of Three Indigenous Small-Sized Greek Tomato (*Lycopersicon esculentum* L.) Cultivars and Four Hybrids of Cherry Tomato under NaCl Salinity Stress. *Commun. Soil Sci. Plant Anal.* **2015**, *46*, 2357–2377. [CrossRef]
36. Grattan, S.R.; Grieve, C.M. Salinity-mineral nutrient relations in horticultural crops. *Sci. Hortic.* **1998**, *78*, 127–157. [CrossRef]
37. Rus, A.; Lee, B.H.; Muñoz-Mayor, A.; Sharkhuu, A.; Miura, K.; Zhu, J.K.; Bressan, R.A.; Hasegawa, P.M. AtHKT1 facilitates Na^+ homeostasis and K^+ nutrition in planta. *Plant Physiol.* **2004**, *136*, 2500–2511. [CrossRef]
38. Shannon, M.; Gronwald, J.; Tal, M. Effects of Salinity on Growth and Accumulation of Organic and Inorganic Ions in Cultivated and Wild Tomato Species. *J. Am. Soc. Hortic. Sci.* **1987**, *112*, 416–423. [CrossRef]
39. Assimakopoulou, A.; Salmas, I.; Roussos, P.A.; Nifakos, K.; Kalogeropoulos, P.; Kostelenos, G. Salt tolerance evaluation of nine indigenous Greek olive cultivars. *J. Plant Nutr.* **2017**, *40*, 1099–1110. [CrossRef]
40. Semiz, G.D.; Suarez, D.L.; Ünlükara, A.; Yurtseven, E. Interactive Effects of Salinity and N on Pepper (*Capsicum annuum* L.) Yield, Water Use Efficiency and Root Zone and Drainage Salinity. *J. Plant Nutr.* **2014**, *37*, 595–610. [CrossRef]

41. Shahid, M.A.; Sarkhosh, A.; Khan, N.; Balal, R.M.; Ali, S.; Rossi, L.; Gómez, C.; Mattson, N.; Nasim, W.; Garcia-Sanchez, F. Insights into the physiological and biochemical impacts of salt stress on plant growth and development. *Agronomy* **2020**, *10*, 938. [CrossRef]
42. Zhang, J.L.; Flowers, T.J.; Wang, S.M. Mechanisms of sodium uptake by roots of higher plants. *Plant Soil* **2010**, *326*, 45–60. [CrossRef]
43. Alfocea, F.P.; Estañ, M.T.; Caro, M.; Bolarín, M.C. Response of tomato cultivars to salinity. *Plant Soil* **1993**, *150*, 203–211. [CrossRef]
44. Peng, Z.; He, S.; Sun, J.; Pan, Z.; Gong, W.; Lu, Y.; Du, X. Na^+ compartmentalization related to salinity stress tolerance in upland cotton (*Gossypium hirsutum*) seedlings. *Sci. Rep.* **2016**, *6*, 34548. [CrossRef] [PubMed]
45. Maathuis, F.J.M.; Ahmad, I.; Patishtan, J. Regulation of Na+ fluxes in plants. *Front. Plant Sci.* **2014**, *5*, 467. [CrossRef] [PubMed]
46. Almeida, D.M.; Margarida Oliveira, M.; Saibo, N.J.M. Regulation of Na+ and K+ homeostasis in plants: Towards improved salt stress tolerance in crop plants. *Genet. Mol. Biol.* **2017**, *40*, 326–345. [CrossRef] [PubMed]
47. Mousavi, A.; Lessani, H.; Babalar, M.; Talaei, A.R.; Fallahi, E. Influence of salinity on chlorophyll, leaf water potential, total soluble sugars, and mineral nutrients in two young olive cultivars. *J. Plant Nutr.* **2008**, *31*, 1906–1916. [CrossRef]
48. Shibli, R.A.; Sawwan, J.; Swaidat, I.; Tahat, M. Increased phosphorus mitigates the adverse effects of salinity in tissue culture. *Commun. Soil Sci. Plant Anal.* **2001**, *32*, 429–440. [CrossRef]
49. Negrão, S.; Schmöckel, S.M.; Tester, M. Evaluating physiological responses of plants to salinity stress. *Ann. Bot.* **2017**, *119*, 1–11. [CrossRef]
50. Tao, R.; Ding, J.; Li, C.; Zhu, X.; Guo, W.; Zhu, M. Evaluating and Screening of Agro-Physiological Indices for Salinity Stress Tolerance in Wheat at the Seedling Stage. *Front. Plant Sci.* **2021**, *12*, 646175. [CrossRef]
51. Anshori, M.F.; Purwoko, B.S.; Dewi, I.S.; Ardie, S.W.; Suwarno, W.B.; Safitri, H. Determination of selection criteria for screening of rice genotypes for salinity tolerance. *Sabrao J. Breed. Genet.* **2018**, *50*, 279–294.
52. Chatzigianni, M.; Ntatsi, G.; Theodorou, M.; Stamatakis, A.; Livieratos, I.; Rouphael, Y.; Savvas, D. Functional Quality, Mineral Composition and Biomass Production in Hydroponic Spiny Chicory (*Cichorium spinosum* L.) Are Modulated Interactively by Ecotype, Salinity and Nitrogen Supply. *Front. Plant Sci.* **2019**, *10*, 1040. [CrossRef]
53. Xu, C.; Mou, B. Evaluation of lettuce genotypes for salinity tolerance. *HortScience* **2015**, *50*, 1441–1446. [CrossRef]
54. Xu, C.; Mou, B. Responses of spinach to salinity and nutrient deficiency in growth, physiology, and nutritional value. *J. Am. Soc. Hortic. Sci.* **2016**, *141*, 12–21. [CrossRef]
55. Xue, X.; Liu, A.; Hua, X. Proline accumulation and transcriptional regulation of proline biothesynthesis and degradation in *Brassica napus*. *BMB Rep.* **2009**, *42*, 28–34. [CrossRef]
56. Žnidarčič, D.; Ban, D.; Šircelj, H. Carotenoid and chlorophyll composition of commonly consumed leafy vegetables in Mediterranean countries. *Food Chem.* **2011**, *129*, 1164–1168. [CrossRef]
57. Zhao, J.; Davis, L.C.; Verpoorte, R. Elicitor signal transduction leading to production of plant secondary metabolites. *Biotechnol. Adv.* **2005**, *23*, 283–333. [CrossRef]
58. Verma, N.; Shukla, S. Impact of various factors responsible for fluctuation in plant secondary metabolites. *J. Appl. Res. Med. Aromat. Plants* **2015**, *2*, 105–113. [CrossRef]
59. Kim, H.J.; Fonseca, J.M.; Choi, J.H.; Kubota, C.; Dae, Y.K. Salt in irrigation water affects the nutritional and visual properties of romaine lettuce (*Lactuca sativa* L.). *J. Agric. Food Chem.* **2008**, *56*, 3772–3776. [CrossRef]
60. Neocleous, D.; Koukounaras, A.; Siomos, A.S.; Vasilakakis, M. Assessing the Salinity Effects on Mineral Composition and Nutritional Quality of Green and Red "Baby" Lettuce. *J. Food Qual.* **2014**, *37*, 1–8. [CrossRef]
61. Garrido, Y.; Tudela, J.A.; Marín, A.; Mestre, T.; Martínez, V.; Gil, M.I. Physiological, phytochemical and structural changes of multi-leaf lettuce caused by salt stress. *J. Sci. Food Agric.* **2014**, *94*, 1592–1599. [CrossRef]
62. Klados, E.; Tzortzakis, N. Effects of substrate and salinity in hydroponically grown *Cichorium spinosum*. *J. Soil Sci. Plant Nutr.* **2014**, *14*, 211–222. [CrossRef]
63. Petropoulos, S.A.; Fernandes, Â.; Dias, M.I.; Pereira, C.; Calhelha, R.C.; Chrysargyris, A.; Tzortzakis, N.; Ivanov, M.; Sokovic, M.D.; Barros, L.; et al. Chemical composition and plant growth of *Centaurea raphanina* subsp. *mixta* plants cultivated under saline conditions. *Molecules* **2020**, *25*, 2204. [CrossRef]
64. Okasha, A.M.; Eldib, E.M.; Elmetwalli, A.H.; Farooque, A.A.; Yaseen, Z.M.; Elsayed, S. Maximization of Water Productivity and Yield of Two Iceberg Lettuce Cultivars in Hydroponic Farming System Using Magnetically Treated Saline Water. *Agriculture* **2022**, *12*, 101. [CrossRef]
65. Carillo, P.; Soteriou, G.A.; Kyriacou, M.C.; Giordano, M.; Raimondi, G.; Napolitano, F.; Di Stasio, E.; Di Mola, I.; Mori, M.; Rouphael, Y. Regulated salinity eustress in a floating hydroponic module of sequentially harvested lettuce modulates phytochemical constitution, plant resilience, and post-harvest nutraceutical quality. *Agronomy* **2021**, *11*, 1040. [CrossRef]
66. Sakamoto, K.; Kogi, M.; Yanagisawa, T. Effects of Salinity and Nutrients in Seawater on Hydroponic Culture of Red Leaf Lettuce. *Environ. Control Biol.* **2014**, *52*, 189–195. [CrossRef]
67. Petropoulos, S.A.; Fernandes, Â.; Dias, M.I.; Pereira, C.; Calhelha, R.; Gioia, F.D.; Tzortzakis, N.; Ivanov, M.; Sokovic, M.; Barros, L.; et al. Wild and cultivated *Centaurea raphanina* subsp. *mixta*: A valuable source of bioactive compounds. *Antioxidants* **2020**, *9*, 314. [CrossRef]
68. Roshdy, A.E.D.; Alebidi, A.; Almutairi, K.; Al-Obeed, R.; Elsabagh, A. The effect of salicylic acid on the performances of salt stressed strawberry plants, enzymes activity, and salt tolerance index. *Agronomy* **2021**, *11*, 775. [CrossRef]

69. Tari, I.; Csiszár, J.; Szalai, G.; Horváth, F.; Pécsváradi, A.; Kiss, G.; Szepesi, Á.; Szabó, M.; Erdei, L. Acclimation of tomato plants to salinity stress after a salicylic acid pre-treatment. *Acta Biol. Szeged.* **2002**, *46*, 55–56.
70. Colla, G.; Kim, H.J.; Kyriacou, M.C.; Rouphael, Y. Nitrate in fruits and vegetables. *Sci. Hortic.* **2018**, *237*, 221–238. [CrossRef]
71. Colonna, E.; Rouphael, Y.; Barbieri, G.; De Pascale, S. Nutritional quality of ten leafy vegetables harvested at two light intensities. *Food Chem.* **2016**, *199*, 702–710. [CrossRef]
72. Kyriacou, M.C.; Rouphael, Y. Towards a new definition of quality for fresh fruits and vegetables. *Sci. Hortic.* **2018**, *234*, 463–469. [CrossRef]
73. Colla, G.; Rouphael, Y.; Cardarelli, M.; Svecova, E.; Rea, E.; Lucini, L. Effects of saline stress on mineral composition, phenolic acids and flavonoids in leaves of artichoke and cardoon genotypes grown in floating system. *J. Sci. Food Agric.* **2013**, *93*, 1119–1127. [CrossRef]
74. Borgognone, D.; Rouphael, Y.; Cardarelli, M.; Lucini, L.; Colla, G. Changes in biomass, mineral composition, and quality of cardoon in response to NO_3^-:Cl^- ratio and nitrate deprivation from the nutrient solution. *Front. Plant Sci.* **2016**, *7*, 978. [CrossRef]
75. Rubinigg, M.; Posthumus, F.; Ferschke, M.; Elzenga, J.T.M.; Stulen, I. Effects of NaCl salinity on 15N-nitrate fluxes and specific root length in the halophyte *Plantago maritima* L. *Plant Soil* **2003**, *250*, 201–213. [CrossRef]

horticulturae

Article

The Effect of Drought on *Sisymbrium officinale* (L.) Wild Species for Potential Cultivation as a Leafy Vegetable

Marta Guarise [1,2], Gigliola Borgonovo [3], Angela Bassoli [3] and Antonio Ferrante [2,*]

1 Agricola 2000 S.c.p.A, Via Trieste 9, 20067 Tribiano, Italy
2 Department of Agriculture and Environmental Science—Production, Landscape, Agroenergy, Università degli Studi di Milano, 20133 Milano, Italy
3 Department of Food, Environmental and Nutritional Science, Università degli Studi di Milano, 20133 Milano, Italy
* Correspondence: antonio.ferrante@unimi.it; Tel.: +39-025-0316-589

Abstract: Leafy vegetables are common components of the human diet and are a source of antioxidant, vitamins, minerals, and bioactive compounds. Fresh-cut or minimally processed industries are always looking for product innovations. Many wild species, based on their composition, can be evaluated as potential vegetables. In this work, hedge mustard has been studied as a potential leafy vegetable, and two wild populations were grown under 100% crop water requirement (WR) and 50% WR. The effect of water reduction was monitored using non-destructive measurements of chlorophyll a fluorescence and by the analytical determination of primary or secondary metabolism associated parameters such as sugars, anthocyanins, carotenoids, phenolic compounds, and nitrate concentrations. The results demonstrated that hedge mustard [*Sisymbrium officinale* (L.) Scop.] can be grown with 50% WR without yield reduction. The yield was not statistically different between the two water regimes and ranged from 22.3 to 40 g plant^{-1} FW. Leaf nitrate concentrations showed high variability in the MI population grown with 100% WR, while in the BG population, they did not change when the WR was shifted from 100% to 50%. The total phenols were 25% higher in the leaves of plants grown under 50% WR in both wild populations. The total sugars and anthocyanins did not show significant variations. Chlorophyll a fluorescence parameters did not show significant changes. The results suggest that hedge mustard can be grown in environments with limited water availability or in the winter season using less water to avoid disease development. The highest yield was obtained from the BG population.

Keywords: antioxidant; chlorophyll; hedge mustard; nitrate; phenolics; water stress; wild species

Citation: Guarise, M.; Borgonovo, G.; Bassoli, A.; Ferrante, A. The Effect of Drought on *Sisymbrium officinale* (L.) Wild Species for Potential Cultivation as a Leafy Vegetable. *Horticulturae* **2023**, *9*, 111. https://doi.org/10.3390/horticulturae9010111

Academic Editor: Qiaomei Wang

Received: 12 December 2022
Revised: 9 January 2023
Accepted: 12 January 2023
Published: 14 January 2023

1. Introduction

Wild herbs have been a source of food or bioactive compounds for humans since ancient times. Wild leafy vegetables are part of the diet of many people around the world. Wild edible species are often underutilized and can be grown to increase the biodiversity of cultivated vegetables and herbs in the human diet [1]. These wild species are collected from nature and used as food. Some herbs have been cultivated to provide an innovation in food production chain. Most wild species in agricultural systems are weeds and can be found in wild lands, along roads, and in urban and peri-urban areas [2]. In cropping systems, they are removed with mechanical or agrochemical treatments. However, the majority of wild species can survive in hostile conditions through the adaptation of their metabolism. These plants can grow by exploiting the limited resources available, such as nutrients and water. These growing conditions increase the accumulation of some defense metabolites that can also have beneficial effects on human health [1]. The screening and selection of these wild species can increase the biodiversity of cultivated vegetables and expand market opportunities for leafy vegetables or baby leaf production. From an agronomic point of view, these species often have a high tolerance against biotic and abiotic stresses, and for this reason, they are very interesting considering the climate change scenario. These traits are also particularly important for reducing fertilizers or agrochemical applications, as requested by the EU Green Deal strategies by 2030 [3].

There are several wild species which have been studied for different proprieties, but just a few have been effectively adopted as crops in horticulture and successfully included in commercial catalogs. Hedge mustard (*Sisymbrium officinale* (L.) Scop.), often referred to by the synonym *Erisymum officinale*, has some traits that could be exploited for the vegetable market [4]. This plant species can be found in Europe, North Africa, and Western Asia. It is an herb famous for its therapeutic properties, which have been used for treating aphonia and hoarseness [5]. For these pharmaceutical proprieties, hedge mustard is also known as "the singers' plant". *Sisymbrium officinale* is commonly used by professional singers as an infuse, or in the form of several herbal preparations such as extracts, tablets, or drops, to prevent or treat voice discomfort [5]. This species belongs to the Brassiceae family, and the main bioactive compounds are represented by glucosinolates and sulfur-containing compounds such as glucoputranjivine, glucojiabutina, napoleiferina, sinalbina, and sinigrin [6]. The concentration of glucosinolates can vary in different plant organs, and glucosinolate chemical species are differentially accumulated in roots, stems, and fruit [7]. These compounds contribute to the antioxidant powder of this species and can have scavenger activity, reducing reactive oxygen species (ROS) and helping plants to adapt and survive in different environments. However, these chemical compounds are also known to have beneficial effects on human health [8]. The successful cultivation of wild species depends on the market demands and the adaptation to agronomic practices. Hedge mustard has been studied as a potential leaf vegetable and its responses to different fertilization levels have been noted [9]. The reduction in fertilization has been found to improve some quality traits, such as pigment and nitrate accumulation. For cultivation purpose, it is very important to improve the germination of seeds. Several studies have been carried out on germination using nitrate and light treatments [10]. Another important parameter that is considered important for cultivated plants is their tolerance to drought.

Therefore, the objective of this work was to evaluate the yield and the way in which quality parameters change when plants are shifted from well-watered conditions to a water reduction regime. The present work is based on the hypothesis that high concentrations of bioactive compounds in this species may be able to confer drought tolerance. The results of this work can be useful for providing agricultural information for growing this species in geographical areas with limited water availability, or during winter, when watering is reduced to avoid disease development.

2. Materials and Methods

2.1. Plant Materials and Water Reduction Management

Two hedge mustard (*Sisymbrium officinale* (L.) Scop.) populations were harvested in the wild near Milan and Bergamo, and were, therefore, been named MI (Milan) and BG (Bergamo), respectively. Plants were grown for seed production and then used for the water reduction experiments [4,9]. Plants were cultivated in greenhouses equipped with sensors for control of the environmental parameters. The greenhouse was located in the Department of Agricultural and Environmental Science of the University of Milan (GPS 45.47688876073164, 9.227201084659612). Seeds were sown in panels filled with commercial fertilized substrate (6 February and 20 February 2018). Seedlings were transplanted into pots when they reached 7–10 cm height (19 April 2018), and cultivation was carried out using complete substrate (Vigorplant, Italy) containing the following components: 13% volcanic peat, 18% calibrated peat, 21% Baltic peat, 22% dark peat, and 26% Irish peat, with pH (H_2O) 6.5–7.5 and electric conductivity (EC) 0.35–0.45 dS m^{-1}. The pH at the beginning of the experiment was 6.5, and plants were placed in plastic pots of 14 cm diameter and 2 L volume.

Ten plants of each pot were fertilized with NPK granular fertilizer (14:7:17) at 4 g/pot, providing 13 g/m^2 N; 7 g/m^2 P_2O_5; 8 g/m^2 K_2O [9].

Water reduction (WR) was performed by gravitropic determination with the full water availability (100%) for the well-watered plants and half that weight of water supplied to the well-water plants (50% WR). During the cultivation period, the pots were weighed in order to restore the same level of water availability that was present at the beginning of the experiment.

The harvest was performed at the 13 BBCH growth stage for each cultivation cycle. Sampling was randomized by casually choosing plants from each pot.

The yields of leaf biomass of the two wild populations were determined to measure total leaf biomass production (3 July 2018). On leaves, the following parameters—anthocyanins, carotenoids, phenols, nitrate, and total sugars—were measured for quality evaluation.

2.2. Chlorophyll *a* Fluorescence

At harvest, the chlorophyll *a* fluorescence was determined using a fluorometer (Handy PEA, Hansatech, United Kingdom). Leaves were covered with leaf clips to fully oxidase the photosystem II. Dark incubation was carried out for 30 min. The chlorophyll a fluorescence induction curve was measured using high-intensity light of 3000 µmol m^{-2} s^{-1} (600 W m^{-2}). Chlorophyll *a* fluorescence-derived parameters were automatically calculated, including the variable fluorescence (Fv) to maximum fluorescence (Fm) and their ratio, Fv/Fm. Induction curve data were used for JIP analyses and the following parameters were reported: performance index (PI), dissipation of energy per reaction center (DIo/RC), and density of reaction centers at the Fm stage (RC/CSm).

2.3. Analytical Determinations

The quality of the hedge mustard populations under water stress conditions was determined. Approximately 1 g of leaves was harvested at the end of the cultivation cycle. Fresh plant matter was determined at the end of the growing cycle. All determinations were performed at harvest.

2.3.1. Chlorophylls and Carotenoids Concentrations

Leaf pigments such as total chlorophylls and total carotenoids were obtained by the extraction of 5 leaf disks of 5 mm diameter, containing 25–40 mg of leaf mix. Extraction was carried out using 5 mL of methanol (99.9%) as solvent, and the disks were kept in a dark, cold room at 4 °C for 24 h. Chlorophyll a, b, and total chlorophylls were immediately determined after extraction. Th readings were performed at 665.2 and 652.4 nm for leaf chlorophylls and 470 nm for total carotenoids. The total chlorophylls and carotenoids were determined by Lichtenthaler's formula [11].

2.3.2. Secondary Metabolites Such as Anthocyanins or Phenolic Compounds Index

Phenolics in leaves were spectrophotometrically determined by direct measurement of the leaf extract. Leaf disks weighing approximately 25–40 mg were extracted using 3 mL methanolic HCl (1%). The mixture was incubated overnight and the supernatant was used for total phenolic determination using the Folin–Ciocalteu method [12]. Data are reported as gallic acid equivalent mg/100 g of fresh weight.

Anthocyanins were spectrophotometrically determined. Leaf disks (5 mm diameter) of approximately 20–30 mg were extracted by 3 mL of methanolic HCl (1%). Mixtures were stored overnight at 4 °C in dark conditions. The anthocyanins were expressed as cyanidin-3-glucoside equivalents spectrophotometrically determined at 535 nm, and quantification was performed using an extinction coefficient (ε) of 29,600 [12].

2.3.3. Leaf Nitrate Concentration

The accumulation of nitrate was spectrophotometrically determined using the salicylic–sulfuric acid method [13]. Nitrate was collected once at the end of the growing cycle when plants reached the growth stage of 13 BBCH.

About 1 g of fresh leaves was ground in 5 mL of distilled water. Extracts were purified using a centrifuge at 4000 rpm for 15 min. The supernatant was used for colorimetric determinations. Samples of 20 µL were taken, as well as 80 µL (5% *w*/*v*) of salicylic acid in concentrated sulfuric acid and 3 mL of NaOH 1.5 N. Samples were cooled, and the absorbance was measured at 410 nm. Nitrate quantification was carried out using to the KNO_3 standard calibration curve.

2.3.4. Total Sugar Determination

The sugar concentration of the fresh leaves was extracted as explained for the determination of nitrate levels. Sugars were determined according the anthrone's assay with slight modification [14]. The reagent (anthrone) was prepared by dissolving 0.1 g of anthrone in 50 mL of 95% H_2SO_4, and it was left for 40 min before use. The extract (200 µL) was transferred to 1 mL of anthrone. Samples were placed on ice for 5 min and then mixed by a vortex. Samples was heated at 95 °C to create the reaction. After 5 min of incubation, samples were cooled, and absorbance was assessed at 620 nm. The total sugar concentration was calculated referring to the glucose standard calibration curve.

2.4. Statistical Analyses

The experimental design was organized as follow: two wild populations (MI and BG); two different water availability treatments, 100% or 50%; ten plants/pots for each water level, for a total number of 20 plants for the MI wild population and 20 for the BG wild population; and two cultivation cycles.

Biological replicates (n = 4) were taken for each water regime level and for each wild population for the measurement of chlorophyll *a* fluorescence, while three biological replicates (n = 3) were taken for each water regime level of each wild population.

Two-way ANOVA (water availability and wild population) was also performed, and differences among the means were determined using Tukey's post-test ($p < 0.05$). The principal component analysis (PCA) was carried out among parameters measured, with eigenvalues > 1.

3. Results

The hedge mustard plants were subjected to water irrigation reduction up to 50% of crop availability. The yield did not significantly change between the two water regimes or the two wild populations. Values ranged from 26 to 40 g plant^{-1} (fresh weight). The MI population grown at 100% WR showed a yield of 26 g plant^{-1} FW and 22.3 g plant^{-1} FW at 50% WR. The BG population showed a yield of 35 g plant^{-1} FW at 100% WR and 40 g plant^{-1} FW at 50% WR. The chlorophyll and carotenoid concentrations did not significantly change according to the different experimental conditions. Chlorophyll values were comprised from 1.6 to 2.1 mg g^{-1} FW (Figure 1A), while total carotenoids ranged from 0.32 to 0.40 mg g^{-1} FW (Figure 1B).

The chlorophyll *a* fluorescence was measured at harvest, and relative parameters were calculated. The results showed that no significant differences were found between wild populations and water regimes. The maximum efficiency values of PSII (Fv/Fm) were from 0.81 to 0.82. The performance index (PI) ranged from 2.02 to 2.52, while the dissipation energy was distributed from 0.46 to 0.57. The reaction centers for the cross-section ranged from 2061 to 2270 (Table 1).

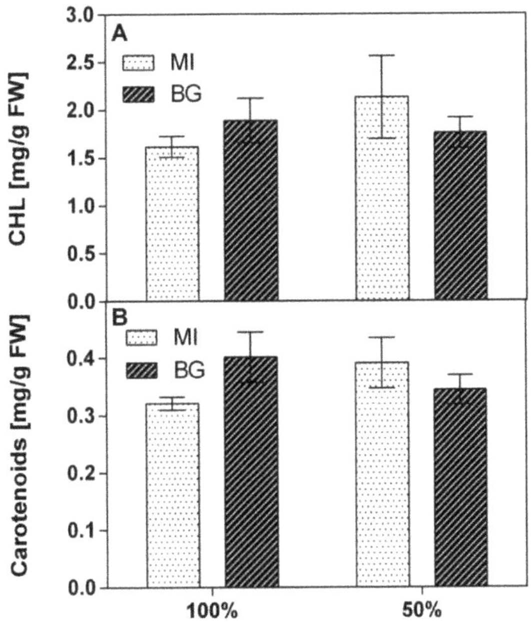

Figure 1. Total chlorophyll (**A**) and total carotenoids (**B**) in fresh leaves of two wild hedge mustard populations, MI and BG, cultivated under different water availability (100% and 50%). Values are means with standard errors (n = 3). Data were subjected to two-way ANOVA, and no significant differences were found.

Table 1. Water reduction (WR): 100% water availability and 50% water reduction and its effect on maximum quantum efficiency (Fv/Fm), performance index (PI), dissipation of energy across reaction centers (DIo/RC), reaction centers per cross-section at Fm (RC/CSm), and electron flux per reaction center (ETo/RC).

WR (%)	Population	Fv/Fm	PI	DIo/RC	RC/CSm	ETo/RC
100	BG	0.82 ± 0.011	2.39 ± 0.453	0.46 ± 0.075	2270.3 ± 202.44	1.3 ± 0.10
100	MI	0.81 ± 0.021	2.14 ± 0.965	0.55 ± 0.153	2073.7 ± 491.53	1.3 ± 0.01
50	BG	0.81 ± 0.015	2.02 ± 0.794	0.57 ± 0.116	2061.2 ± 434.31	1.2 ± 0.11
50	MI	0.82 ± 0.020	2.52 ± 0.799	0.50 ± 0.123	2223.3 ± 403.03	1.3 ± 0.11

Values are means with standard deviations. Data were subjected to two-way ANOVA, but differences were not significant.

The electron flux per reaction center (ETo/RC) is an index that measures the flux of electrons transferred to the photosynthesis machinery, and it is closely associated with the photosynthetic activity of plants. Data measured indicated that there were no significant differences, and values were of 1.2–1.3 a.u. (Table 1).

Of the two populations grown with limited water availability, the nitrate concentration in the leaves was higher in MI compared to BG in normal conditions, with values that ranged from 3200 to 5800 mg/kg FW. Under limited water availability, both populations showed the same values. However, the BG population did not show any changes in concentration between normal and drought conditions, while the MI population showed a nitrate reduction (Figure 2A).

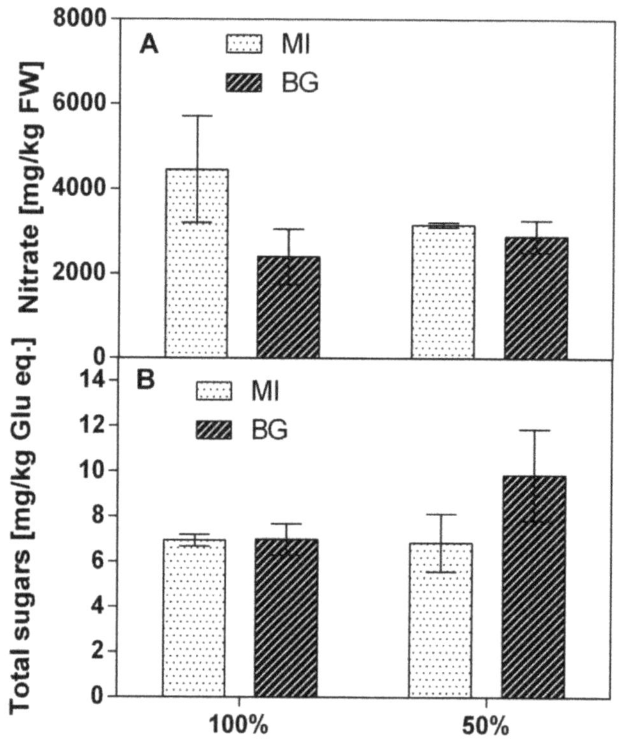

Figure 2. Nitrate (**A**) and total sugar (**B**) content in fresh leaves of two wild hedge mustard populations, MI and BG, cultivated under different water availability levels (100% and 50%). Values are means with standard errors (n = 3). Data were subjected to two-way ANOVA, and no significant differences were found.

The total sugars did not significantly change between water availability regimes and populations. The concentration ranged from 7 to 10 mg kg^{-1} FW on average, even if in the BG population, data showed a wider variability with higher means at 50% water availability (Figure 2B).

The effect of water stress was evaluated on the basis of the secondary metabolism compound changes. Anthocyanin concentrations ranged from 41 to 47 mg 100^{-1} FW, and differences were not significant between water regimes and wild populations (Figure 3A). Phenolic compounds significantly increased under 50% water reduction in both populations. Statistical analysis revealed that the interaction of WR $\times$ Population was not significant ($p = 0.82$), while the WR factor was significant, at $p = 0.0006$. The population factor, instead, was not significant, with a $p = 0.23$. At 100% water availability, the phenolic compounds ranged from 164 to 174 mg 100 g^{-1} FW, while at 50%, they ranged from 217 to 233 mg 100 g^{-1} FW (Figure 3B). In both populations, the reduction in water induced an increase of 25% in total phenolics.

Significant correlations were found among chlorophyll and chlorophyll a fluorescence parameters, anthocyanins, and phenols. Anthocyanins were also significantly correlated with the number of reaction centers, the electron flux, and dissipation energy through the reaction centers (Table 2).

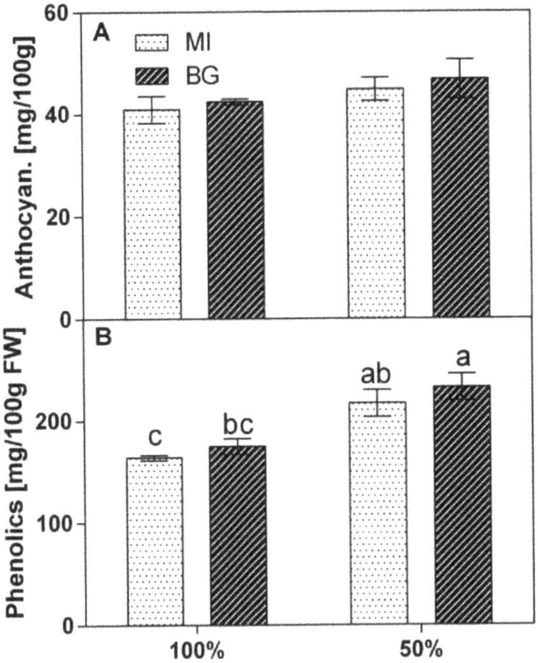

Figure 3. Anthocyanins expressed cyanidin-3-glucosides mg/100g FW (**A**), and phenolic compounds expressed as gallic acid mg/100 g FW (**B**). Their content in the fresh leaves of two wild hedge mustard populations, MI and BG, cultivated under different water availability levels (100% and 50%), were determined. Values are means with standard errors (n = 3). Data were subjected to two-way ANOVA, and significant differences among means were determined using Tukey's post-test. Different letters indicate significant differences ($p < 0.05$).

Table 2. Correlation coefficients of all parameters measured and significant correlations is reported with * for $p < 0.05$.

	Total Sugars	Nitrate	Total Phenols	Anthocyanins	Chl	Carotenoids	PI	Fv/Fm	DIo/RC	RC/CSm	ETo/RC
Total sugars		−0.016	0.215	0.113	0.090	−0.037	−0.194	−0.226	0.159	−0.100	0.194
Nitrate	−0.016		−0.223	−0.192	−0.206	−0.337	−0.342	−0.231	0.292	−0.341	0.171
Total phenols	0.215	−0.223		0.506 *	0.210	0.059	−0.037	−0.133	0.143	−0.051	0.327
Anthocyanins	0.113	−0.192	0.506 *		0.294	0.230	−0.483	−0.447	0.541 *	−0.535 *	0.619 *
Chl	0.090	−0.206	0.210	0.294		0.881 *	−0.0922	−0.096	0.095	−0.124	0.213
Carotenoids	−0.037	−0.337	0.059	0.230	0.881 *		−0.109	−0.064	0.031	−0.113	0.038
PI	−0.194	−0.342	−0.037	−0.483	−0.092	−0.109		0.949 *	−0.954 *	0.977 *	−0.673 *
Fv/Fm	−0.226	−0.231	−0.133	−0.447	−0.096	−0.065	0.949 *		−0.953 *	0.947 *	−0.670 *
DIo/RC	0.159	0.292	0.143	0.541 *	0.095	0.031	−0.954 *	−0.954 *		−0.967 *	0.827 *
RC/CSm	−0.100	−0.342	−0.052	−0.535 *	−0.124	−0.113	0.977 *	0.947 *	−0.967 *		−0.702 *
ETo/RC	0.194	0.171	0.327	0.620 *	0.213	0.038	−0.673 *	−0.670 *	0.827 *	−0.702 *	

The PCA performed using all analytical data revealed that there was no clear separation of components among the parameters measured (Figure 4). The results indicated that there was no significant separation between the wild population and WR regimes.

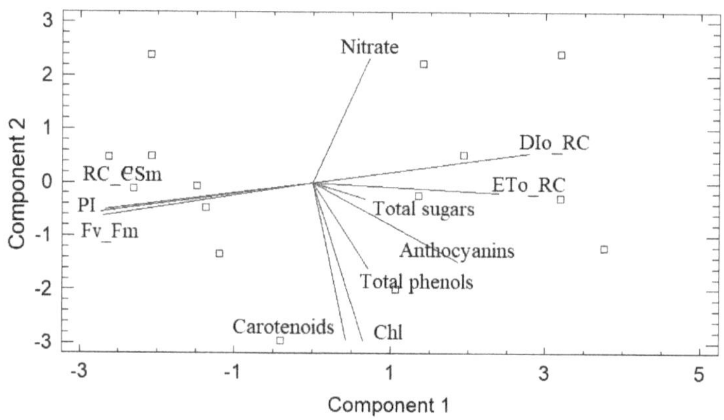

Figure 4. Biplot of principal component analysis results of all parameters measured in the two populations under two water regimes.

4. Discussion

In the cultivation of leafy vegetables, the water availability for fresh-cut or minimally processed production chains is very important. The reduction in water supply can prevent excessive leaching and loss of nutrients into the soil, avoiding underground water pollution. At the agronomic level, during winter growing cycles, the reduction in the irrigation water can limit the incidence of fungal diseases. However, the reduction in water supply can increase the nitrate concentration, with a negative effect on quality [15]. In the present study, the yield of hedge mustard was different between populations, but it was not affected by water reduction, indicating the ability of this species to grow under water shortage. Wild species are usually more tolerant than their cultivated relatives. This evidence has been reported for cotton, potato, maize, rice, and soybean [16,17] plants. It is well-known that wild relatives are a good source of tolerant traits against abiotic stresses, which is useful in genetic improvement programs. However, transferring a trait from a wild species to cultivated ones is not easy and requires many years of work. Therefore, direct screening of suitable new wild species for agricultural purposes can provide a fast product innovation. In the fresh-cut industry, the identification of new crops with high nutritional quality is highly desired [18]. Several studies have been carried out to provide information on the potential use of wild or underutilized species as baby leaves for the fresh-cut industries. The effect of water reduction availability was evaluated on both the primary and secondary metabolism. The influence of water limitation on the primary metabolism was evaluated at harvesting time by measuring the chlorophyll concentration and the chlorophyll *a* fluorescence [19]. The chlorophyll concentration is important in leafy vegetables because it is connected to the photosynthetic machinery and the light harvesting complex, but also to the visual appearance of the produce, along with anthocyanins [20]. Chlorophyll *a* fluorescence measurements are important for understanding the stress conditions of crops, and they allow us to estimate the light use efficiency [21]. The obtained results were similar to those reported in previous studies focused on the comparison of hedge mustard in fertilization experiments [9] and wild populations [4]. The total carotenoids showed the same trend as chlorophyll, and did not change. Carotenoids are antioxidants, and have chlorophyll protection functions. However, they can contribute to the production of total antioxidants and enhance the nutritional quality of the product.

The nitrate concentration was measured since, for leafy vegetable production, there are limits imposed by European Union [22] for their commercialization. The EU regulation n. 1258/2011 reported that for some Brassiceae, including the *Sisymbrium tenuifolium*, the limits were differentiated among crops and in different cultivation periods. In winter (1 October to 31 March) the limits are 7000 mg kg^{-1} FW, and 6000 mg kg^{-1} FW in summer

(1 April to 30 September). The results demonstrated that the two populations have different nitrate accumulation abilities, but neither nitrate concentrations overcame the EU limits under water stress. The MI showed higher leaf nitrate concentrations, and these findings confirm previous data [9]. However, the leaf pigment concentration can vary with growing periods [4,9]. Nitrates, as sugars, can contribute to osmotic adjustment under drought stress. A slight increase in total sugars was observed under 50% water reduction conditions in BG population. The lack of increase in these two parameters suggests that the 50% water reduction did not induce significant stress in the hedge mustard.

Total sugars are important, because they represent the energy source of plants for their metabolism and are required for respiration after harvest. The amount of total sugars is important for the post-harvest storage duration and shelf-life of leafy vegetables.

Among the secondary metabolites, anthocyanins and phenolic compounds were monitored, and the results indicated that lower water availability increased the phenolic compounds. Since glucosinolates and phenolic compounds contribute to beneficial effects on human health, many of the secondary metabolites have pharmacological proprieties [8]. *Sisymbrium officinale* (L.) has been widely studied for its potential pharmaceutical applications. It has been found that its extracts are able to inhibit mutagenicity in vitro [23]. The isopropylisothiocyanate and 2-buthylisothiocyanate isolated from hedge mustard were tested in vitro and found to be potent agonists of TRPA1 [24]. Glucoputranjivin has been also found to be a selective agonist of the T2R16 receptor [25]. Anti-arthritic activity was also observed for the dichloromethane extracts in vitro using rat liver microsomal cells. *S. officinale* extracts were able to reduce the production of the pro-inflammatory mediator nitric oxide, as well as lipid peroxidation [6]. The efficacy of *S. officinale* extracts depends on the concentration of bioactive compounds. The obtained results suggest that a water reduction of up to 50% of the water availability does not affect the yield, and even induces an increase in bioactive compounds of this species. Similar results have been reported for rapeseed (*Brassica napus* L.), which, grown under drought stress, showed an increase in chlorophylls, carotenoids, total pigment, phenolic compounds, flavonoids, and antioxidant activities [26]. The effects of water reduction was also studied in wild rocket (*Diplotaxis tenuifolia* L.) for which the water supply was 50% of the evapotranspiration, and it was found to increase total phenols, total carotenoids, and nitrates [15].

The increase in antioxidant compounds under abiotic stress is a crop defense mechanism that invests energy for the biosynthesis of secondary metabolites. Antioxidant compounds reduce the ROS accumulation and increase crop tolerance to abiotic stress. Crops with fast and positive responses to the abiotic stress have a higher adaptation ability and a higher nutritional quality.

5. Conclusions

Hedge mustard could be considered a potential cultivable wild species as a new leafy vegetable for the fresh-cut industry production chain. It can be suitable for expanding the vegetable production in geographical areas with reduced water availability. This species could be an optimal crop for winter cultivation, when irrigation is reduced to avoid the incidence of fungal diseases. Between the two wild populations, the best performance was observed in the BG population. However, further investigations are required for understanding the minimum water availability that affects the crop's performance, in order to better exploit its tolerance against drought.

Author Contributions: Conceptualization, A.B. and A.F.; methodology, M.G. and G.B.; formal analysis, M.G. and G.B.; investigation, M.G.; data curation, M.G.; writing—original draft preparation, M.G. and A.F.; writing—review and editing, G.B., A.B. and A.F.; visualization, M.G.; supervision, A.F.; project administration, A.B.; funding acquisition, A.B. All authors have read and agreed to the published version of the manuscript.

Funding: This research was funded by Fondazione Cariplo, ERISIMO A MILANO project, no. 2017-1653.

Data Availability Statement: Data are available upon request.

Conflicts of Interest: The authors declare no conflict of interest.

References

1. Sánchez-Mata, M.C.; Cabrera Loera, R.D.; Morales, P.; Fernández-Ruiz, V.; Cámara, M.; Díez Marqués, C.; Tardío, J. Wild vegetables of the Mediterranean area as valuable sources of bioactive compounds. *Genet. Resour. Crop Evol.* **2012**, *59*, 431–443. [CrossRef]
2. Zorzan, M.; Zucca, P.; Collazuol, D.; Peddio, S.; Rescigno, A.; Pezzani, R. *Sisymbrium officinale*, the plant of singers: A review of its properties and uses. *Planta Medica* **2020**, *86*, 307–311. [CrossRef]
3. Montanarella, L.; Panagos, P. The relevance of sustainable soil management within the European Green Deal. *Land Use Policy* **2021**, *100*, 104950. [CrossRef]
4. Guarise, M.; Borgonovo, G.; Bassoli, A.; Ferrante, A. Evaluation of two wild populations of hedge mustard (*Sisymbrium officinale* (L.) Scop.) as a potential leafy vegetable. *Horticulturae* **2019**, *5*, 13. [CrossRef]
5. Calcinoni, O.; Borgonovo, G.; Cassanelli, A.; Banfi, E.; Bassoli, A. Herbs for Voice Database: Developing a rational approach to the study of herbal remedies used in voice care. *J. Voice* **2021**, *35*, 807-e33. [CrossRef]
6. Amodeo, V.; Marrelli, M.; Pontieri, V.; Cassano, R.; Trombino, S.; Conforti, F.; Statti, G. *Chenopodium album* L. and *Sisymbrium officinale* (L.) Scop.: Phytochemical Content and in vitro antioxidant and anti-Inflammatory potential. *Plants* **2019**, *8*, 505. [CrossRef]
7. Đulović, A.; Popović, M.; Burčul, F.; Čikeš Čulić, V.; Marijan, S.; Ruščić, M.; Anđelković, N.; Blažević, I. Glucosinolates of Sisymbrium officinale and S. orientale. *Molecules* **2022**, *27*, 8431. [CrossRef] [PubMed]
8. Carnat, A.; Fraisse, D.; Carnat, A.P.; Groubert, A.; Lamaison, J.L. Normalization of hedge mustard, *Sisymbrium officinale* L. *Ann. Pharm. Françaises* **1998**, *56*, 36–39.
9. Guarise, M.; Borgonovo, G.; Bassoli, A.; Ferrante, A. Effect of fertilization on yield and quality of *Sisymbrium officinale* (L.) Scop. grown as leafy vegetable crop. *Agronomy* **2019**, *9*, 401. [CrossRef]
10. Hilhorst, H.W.; Karssen, C.M. Nitrate reductase independent stimulation of seed germination in *Sisymbrium officinale* L. (hedge mustard) by light and nitrate. *Ann. Bot.* **1989**, *63*, 131–137. [CrossRef]
11. Lichtenthaler, H.K. Chlorophylls and carotenoids: Pigments of photosynthetic membranes. *Methods Enzymol.* **1987**, *148*, 350–382.
12. Ferrante, A.; Incrocci, L.; Maggini, R.; Serra, G.; Tognoni, F. Colour changes of fresh-cut leafy vegetables during storage. *J. Food Agric. Environ.* **2004**, *2*, 40–44.
13. Cataldo, C.A.; Maroon, M.; Schrader, L.E.; Youngs, V.L. Rapid colorimetric determination of nitrate in plant tissue by titration of salicylic acid. *Commun. Soil Sci. Plant Anal.* **1975**, *6*, 71–80. [CrossRef]
14. Yemm, E.W.; Willis, A.J. The estimation of carbohydrates in plant extracts by anthrone. *Biochem. J.* **1954**, *57*, 508–514. [CrossRef]
15. Schiattone, M.I.; Boari, F.; Cantore, V.; Castronuovo, D.; Denora, M.; Di Venere, D.; Perniola, M.; Sergio, L.; Todorovic, M.; Candido, V. Effect of water regime, nitrogen level and biostimulants application on yield and quality traits of wild rocket [*Diplotaxis tenuifolia* (L.) DC.]. *Agric. Water Manag.* **2023**, *277*, 108078. [CrossRef]
16. Arvin, M.J.; Donnelly, D.J. Screening potato cultivars and wild species to abiotic stresses using an electrolyte leakage bioassay. *J. Agric. Sci. Technol.* **2008**, *10*, 33–42.
17. Mammadov, J.; Buyyarapu, R.; Guttikonda, S.K.; Parliament, K.; Abdurakhmonov, I.Y.; Kumpatla, S.P. Wild relatives of maize, rice, cotton, and soybean: Treasure troves for tolerance to biotic and abiotic stresses. *Front. Plant Sci.* **2018**, *9*, 886. [CrossRef]
18. Nicola, S.; Cocetta, G.; Ferrante, A.; Ertani, A. Fresh-cut produce quality: Implications for postharvest. In *Postharvest Handling*; Academic Press: Cambridge, MA, USA, 2022; pp. 187–250.
19. Stirbet, A. On the relation between the Kautsky effect (chlorophyll a fluorescence induction) and photosystem II: Basics and applications of the OJIP fluorescence transient. *J. Photochem. Photobiol. B Biol.* **2011**, *104*, 236–257. [CrossRef]
20. Grieco, M.; Tikkanen, M.; Paakkarinen, V.; Kangasjärvi, S.; Aro, E.M. Steady-state phosphorylation of light-harvesting complex II proteins preserves photosystem I under fluctuating white light. *Plant Physiol.* **2012**, *160*, 1896–1910. [CrossRef]
21. Maxwell, K.; Johnson, G.N. Chlorophyll fluorescence—A practical guide. *Ex. Bot. J.* **2000**, *51*, 659–668. [CrossRef]
22. Cavaiuolo, M.; Ferrante, A. Nitrates and glucosinolates as strong determinants of the nutritional quality in rocket leafy salads. *Nutrients* **2014**, *6*, 1519–1538. [CrossRef] [PubMed]
23. Di Sotto, A.; Di Giacomo, S.; Toniolo, C.; Nicoletti, M.; Mazzanti, G. *Sisymbrium officinale* (L.) Scop. and its polyphenolic fractions inhibit the mutagenicity of Tert-butylhydroperoxide in *Escherichia Coli* WP2*uvr*AR strain. *Phytother. Res.* **2016**, *30*, 829–834. [CrossRef] [PubMed]
24. Borgonovo, G.; Zimbaldi, N.; Guarise, M.; De Nisi, P.; De Petrocellis, L.; Schiano Moriello, A.; Bassoli, A. Isothiocyanates and glucosinolates from *Sisymbrium officinale* (L.) Scop.("the Singers' Plant"): Isolation and in vitro assays on the somatosensory and pain receptor TRPA1 channel. *Molecules* **2019**, *24*, 949. [CrossRef] [PubMed]

25. Borgonovo, G.; Zimbaldi, N.; Guarise, M.; Bedussi, F.; Winnig, M.; Vennegeerts, T.; Bassoli, A. Glucosinolates in *Sisymbrium officinale* (L.) Scop.: Comparative analysis in cultivated and wild plants and in vitro assays with T2Rs bitter taste receptors. *Molecules* **2019**, *24*, 4572. [CrossRef]
26. Salami, M.; Heidari, B.; Tan, H. Comparative profiling of polyphenols and antioxidants and analysis of antiglycation activities in rapeseed (*Brassica napus* L.) under different moisture regimes. *Food Chem.* **2023**, *399*, 133946. [CrossRef]

horticulturae

Article

Domestication of Wild Edible Species: The Response of *Scolymus hispanicus* Plants to Different Fertigation Regimes

Beatriz H. Paschoalinotto [1,2,†], Nikolaos Polyzos [3,†], Maria Compocholi [3], Youssef Rouphael [4], Alexios Alexopoulos [5], Maria Inês Dias [1,2], Lillian Barros [1,2,*] and Spyridon A. Petropoulos [3,*]

1 Centro de Investigação de Montanha (CIMO), Instituto Politécnico de Bragança, Campus de Santa Apolónia, 5300-253 Braganca, Portugal
2 Laboratório Associado para a Sustentabilidade e Tecnologia em Regiões de Montanha (SusTEC), Instituto Politécnico de Bragança, Campus de Santa Apolónia, 5300-253 Braganca, Portugal
3 Department of Agriculture Crop Production and Rural Environment, University of Thessaly, 38446 Volos, Greece
4 Department of Agricultural Sciences, University of Naples Federico II, 80055 Portici, Italy
5 Laboratory of Agronomy, Department of Agriculture, University of the Peloponnese, Antikalamos, 24100 Kalamata, Greece
* Correspondence: lillian@ipb.pt (L.B.); spetropoulos@uth.gr (S.A.P.)
† These authors contributed equally to this work.

Citation: Paschoalinotto, B.H.; Polyzos, N.; Compocholi, M.; Rouphael, Y.; Alexopoulos, A.; Dias, M.I.; Barros, L.; Petropoulos, S.A. Domestication of Wild Edible Species: The Response of *Scolymus hispanicus* Plants to Different Fertigation Regimes. *Horticulturae* **2023**, *9*, 103. https://doi.org/10.3390/horticulturae9010103

Academic Editor: Rita Maggini

Received: 11 December 2022
Revised: 6 January 2023
Accepted: 10 January 2023
Published: 12 January 2023

Abstract: *Scolymus hispanicus* L. is a wild edible species with wide distribution in the Mediterranean area. Recent research has focused on the domestication of wild edible greens, which is essential for the preservation of agroecosystems and the increase in biodiversity, especially under the adversely changing climate conditions. In the present work, the aim was to evaluate the response of *S. hispanicus* plants to different fertilization regimes that varied in the amounts of nitrogen, phosphorus and potassium in regard to plant growth and chemical composition of leaves. For this purpose, plants were grown in pots within an unheated greenhouse. Seven experimental treatments were used, including six fertigation regimes (SH1-SH6) and the control treatment (SHC), where no fertilizers were added. Fresh yield was beneficially affected by the treatments that included a high content of P and K (e.g., SH3 and SH5), while lesser amounts of these macronutrients (e.g., SH1 and SH4) resulted in higher chlorophyll content (SPAD index) and leaf area. In terms of mineral profile, high amounts of P and K improved dietary fiber and carbohydrates content, whereas the untreated plants had the highest content of ash, fat and crude protein. Oxalic and quinic acid were the major organic acids detected, with fertigation regimes significantly reducing their content compared to the control treatment. α-tocopherol was the only isoform of vitamin E detected in all the samples, while glucose and fructose were the most abundant sugars, with their highest content detected in control and SH4 treatments, respectively. *Scolymus hispanicus* leaves were rich in macro and micro minerals, while their contents varied depending on the fertigation regime. Finally, α-linolenic, palmitic, and linoleic acid were the major fatty acids detected, while their contents were beneficially affected by low nutrient inputs (e.g., untreated plants and SH1 and SH2 treatments). In conclusion, the regulation of nutrient solution seems to be an effective practice to increase fresh yield in *S. hispanicus* without compromising the nutritional profile of the edible product, while low inputs of macronutrients such as P and K may improve the chemical composition of the species, especially in terms of n-fatty acids.

Keywords: common golden thistle; nutritional value; mineral profile; chemical composition; wild edible greens; Mediterranean diet; Spanish oyster thistle; organic acids; tocopherols

1. Introduction

Modern farming systems aim to obtain maximum yields to ensure the alarmingly increasing food demands [1,2], while at the same time, the anthropogenic activities tend to gradually reduce the available arable land [3]. Therefore, a new approach is needed focusing not only on intensification of cropping systems, but also on maximizing the efficient use of the available arable land, including degraded soils that cannot be grown with conventional crops. In this context, the valorization of underutilized and neglected species seems to be a promising alternative to the existing crops, especially when considering the ongoing climate crisis that severely affects conventional crop production [4], which may provide new sources of food with improved nutritional properties [5]. Moreover, the integration of such species in farming systems ensures the preservation of agrobiodiversity heritage and reduces the risk of genetic erosion in agroecosystems due to intensified cropping systems [1,6–8].

Scolymus hispanicus L. (also known as common golden thistle or Spanish oyster thistle) is a wild annual or perennial herb belonging to the Asteraceae family with wide distribution in the Mediterranean basin [9–11]. It is usually found in agricultural ecosystems and is considered a difficult-to-control and noxious weed [12,13]. However, although it is undesirable in commercial farms, it is highly appreciated as a wild edible green due to its high nutritional value and beneficial health effects, especially in the countries of southern Europe [9,14]. It is a common ingredient in various gourmet and local dishes of the so-called Mediterranean diet where it is consumed in raw, boiled or fried form [9,11,15,16]. The most commonly consumed plant parts are the flowers, midribs and petioles of leaves, as well as the cortex of the roots, which after post-harvest processing can be used in various dishes [10,17]. According to Disciglio et al. [18], wild *S. hispanicus* plants are rich sources of Mg and Ca and contain low amounts of nitrates, while Rubio et al. [19,20] identified various flavonoids and phenolic acids. Petropoulos et al. [21] suggested luteolin and kaempferol glucuronides as the major phenolic compounds, while Vardavas et al. [22] detected moderate amounts of vitamins (K1 and C), carotenoids (lutein and carotene), Tbatou et al. [23] detected high dietary fiber content, and Morales et al. [24] detected low amounts of α- and total tocopherol. Moreover, Vardavas et al. [25] reported a balanced content of n-6 and n-3 fatty acids (a ratio of 1.06) and high amounts of palmitic, linoleic and α-linolenic acids.

Considering consumers' awareness regarding the origin of food and the production practices implemented, especially regarding the inputs of agrochemicals, the use of sustainable means for crop production is essential for fulfilling the market demands [26]. The introduction of alternative crops such as the various wild and underutilized species falls within this context due to their low requirements in agrochemicals and natural resources (e.g., fertilizers and water) and their efficient adaption strategies to various abiotic and biotic stressors [27,28]. Moreover, the commercialization of these species is pivotal for the reduction of genetic erosion risks related to irrational harvest and anthropogenic activities [29]. During the last few years, several studies have focused on the chemical characterization and the bioactivities of various wild edible plants [30–32], while numerous ethnopharmacological studies have highlighted their contemporary uses in modern diets and their positive health effects [13,22,33–38]. However, in order to domesticate these species, useful information regarding the best practice guides that farmers should follow in order to achieve high yields and high quality of the final product should be also provided [9,13]. Thus far, various species have been suggested for commercial exploitation in small-scale farming systems of the Mediterranean, including *Cichorium spinosum* [39–41], *Portulaca oleracea* [42–44], *Sanguisorba minor* [45,46] *Crithmum maritimum* [47–49] and several others [21,31,50,51]. Among the cultivation practices, the fertilization regime has a significant impact on the yield, the chemical composition and bioactive properties, and the optimum fertilization has to be considered for the commercial production of final products with similar quality as the wild counterparts [3,52–55]. Moreover, the existing genotypic diversity among the numerous ecotypes of these species suggests a wide variation in chemical composition, which along with the effect of growing condition, may result in significant differences in the chemical profile of wild edible greens [56,57].

Despite the prolific studies regarding the chemical properties and the cultivation practices of various wild edible species, there is scarce literature for *Scolymus hispanicus* since most of the studies focus on the chemical characterization and bioactivities of plants collected from the wild [18,36,58]. On the other hand, Papadimitriou et al. [59] suggested that *S. hispanicus* is moderately tolerant to salinity species that could be utilized in saline agriculture, while they also suggested its introduction in soilless cropping systems with reduced macronutrients requirements [60]. Considering the limited information about *S. hispanicus* cultivation, the aim of the present study was to evaluate the impact of different fertilization regimes that varied in the amounts of the main macronutrients (e.g., N, P, K) on the growth, nutritional and mineral profile and chemical composition of *S. hispanicus* plants. The presented results will be helpful for the integration of the species as an innovative crop in the existing farming systems, especially in the small-scale farms of the Mediterranean, while they provide a best-practice guide for the fertilization of the species, focusing on high yields without compromising the quality of the final product.

2. Materials and Methods

2.1. Plant Material and Growing Conditions

The trial was conducted at the experimental farm of the University of Thessaly in Velestino (Greece; 39°37′18.6″ N, 22°22′55.1″ E) during the growing period of October 2020–April 2021. The experiment was conducted in the unheated glasshouse of the experimental field. Seeds of *Scolymus hispanicus* were sown in seed trays on October 2021, and young seedlings were transplanted in 6 L plastic pots with peat (Klassman–Deilmann KTS2, Geeste, Germany) and perlite (1:1, *v*/*v*) in January 2021, when the plants had formed 3–4 true leaves. Physicochemical properties of peat were as follows: bulk density 0.12 g/cm^3, water holding capacity 218.5%, pH 6.0, electrical conductivity (EC) 0.35 dS/m, organic matter 47.5%, carbon 27.5%, nitrogen 0.14%, C/N 196.8, P 160 mg/L, and K (cmol/kg) 46.03 [51]. Mean air temperature throughout the experimental period was as follows: October, 16.4 °C; November, 10.7 °C; December, 7.8 °C, January, 6.5 °C, February, 8.6 °C; March 12.4 °C; and April 16.7 °C. Data for temperatures inside the greenhouse were obtained from Onset HOBO RH/Temp data logger (Onset Computer Corporation, MA, USA) [61]. The tested fertigation treatments have already been described in detail in a similar study by our team regarding the commercial cultivation of *Cichorium spinosum* [39,52]. In brief, the applied treatments differed in the amounts of N:P:K, e.g., 100:100:100 (SH1), 200:100:100 (SH2), 200:200:200 (SH3), 300:100:100 (SH4), 300:200:200 (SH5) and 300:300:300 ppm (SH6) of N:P:K, as well as the control treatment without the addition of fertilizers (SHC). Stock solutions were prepared with Atlas 20 -20–20 (4.8% ammonium N, 5.0% nitric N, 10.2% ureic N; 20% P2O5; 20% K2O)+ TE (trace elements) fertilizer (Gavriel; S.A., Volos, Greece) for the preparation of 100:100:100 ppm (SH1), 200:200:200 ppm (SH3), and 300:300:300 ppm (SH6), while for the rest of the solutions (200:100:100 ppm (SH2), 300:100:100 ppm (SH4), 300:200:200 ppm (SH5)), the extra amount of nitrogen was achieved with the addition of ammonium nitrate fertilizer (34.5% of N; Gavriel; S.A., Volos, Greece). The control treatment included tap water with no fertilizers added (SHC) [39,52]. The application of treatments was performed manually once or twice per week via the abovementioned solutions, while each treatment included fifteen plants (n = 15) with one plant per pot and 105 plants in total. The total amount of nutrient solution for all the treatments was 1.8 L. Therefore, the abovementioned treatments refer to the following amounts of N:P:K per hectare (assuming a soil depth of 0,15 m as the pot height): SH1, 135-135-135 kg of N:P:K per hectare; SH2, 270-135-135 kg of N:P:K per hectare; SH3, 270-270-270 kg of N:P:K per hectare; SH4, 405-135-135 kg of N:P:K per hectare; SH5, 405-270-135 kg of N:P:K per hectare; SH6, 405-405-405 kg of N:P:K per hectare. The experiment was laid out according to completely randomized design (CRD).

Before harvest, chlorophyll content of leaves (SPAD values) was recorded from 10 plants from each treatment. Harvest took place on April 9, 2021, and morphological traits such as the weight of plant (g), the number of leaves/plant, the weight of leaves/plant (g), the dry matter of leaves (%), the total leaf area (cm^2) and specific leaf area (m^2/kg) were also determined. Dry matter content of leaves, leaf area as well as specific leaf area were calculated from five plants from each treatment. Dry matter content was measured after drying fresh samples of leaves in a forced-air oven at 72 °C to constant weight. Leaf area was measured with a leaf area meter (LI-3100C, LICOR Biosciences; Hellamco S.A., Greece). Specific leaf area was calculated by dividing the dry weight of leaves of each plant by the corresponding leaf area.

2.2. Chemical Analysis

2.2.1. Nutritional Profile

The proximate composition of the edible leaves was performed according to the protocols of the Association of Official Analytical Chemists (crude protein: AOAC, 991.02; total fat: AOAC, 989.05; total dietary fiber: AOAC, 991.43 and 992.16; ash: AOAC, 935.42; and carbohydrates (by difference)) [62]. The results were expressed in g/100 g fw (fresh weight). The energy value was calculated using the formula: Energy = 4 × (protein + carbohydrate) + 2 × (total dietary fiber) + 9 × (total fat), and the results were expressed in kcal/100 g dw.

2.2.2. Organic Acids

Organic acids were analyzed in the dry powder of *S. hispanicus* by ultrafast liquid chromatography coupled to a photodiode array detector programmed to record at 215 nm as the preferred wavelength (UFLC-PDA; Shimadzu Corporation, Kyoto, Japan), and a C_{18} reverse phase column (250 × 4.6 mm, 5 μm, Phenomenex; Torrance, CA, USA) was used for compound separation at 35 °C [63]. The results were expressed in mg per 100 g of dry weight.

2.2.3. Tocopherols

The analysis of tocopherols was carried out based on the protocols described by Barros et al. [63], using a high performance liquid chromatography coupled to a fluorescence detector with a Polyamide II normal phase column (250 mm × 4.6 mm, 5 μm) at 35 °C (HPLC-FP, Knauer, Smartline system 1000, Berlin, Germany). The results were expressed in mg per 100 g of dry weight.

2.2.4. Free Sugars

Free sugars were determined with high-performance liquid chromatography coupled to a refractive index detector and a 100-5 NH_2 Eurospher column (HPLC-RI, Knauer, Smartline system 1000, Berlin, Germany) as described by Barros et al. [63]. The results were expressed in g per 100 g of dry weight of the plant.

2.2.5. Mineral Content

The mineral composition was determined according to the respective AOAC protocol [62]. In brief, the dry powder of *S. hispanicus* was digested with 10 mL of nitric acid in a microwave extraction system at 200 °C and 1600 watts for 30 min and then made up to a final volume of 50 mL with distilled water. The mineral content in terms of potassium (K), sodium (Na), calcium (Ca), magnesium (Mg), iron (Fe), manganese (Mn), zinc (Zn) and copper (Co) was determined through atomic absorption spectrophotometry (Perkin Elmer 1100B, Waltham, MA, USA).

2.2.6. Fatty Acids

Fatty acid methyl esters (FAME) were determined after trans-esterification of the lipid fraction obtained from the dry powder samples of *S. hispanicus* through Soxhlet extraction, following the protocol described by Barros et al. [63]. The results were expressed as relative percentage (%).

2.3. Statistical Analysis

Plant growth measurements were performed in 15 plants (n = 15) per treatment, except for SPAD index (n = 10), and dry matter content, leaf area and specific leaf area (n = 5). Chemical analyses were performed in triplicate in three batch samples of each treatment.

All data were checked for normal distribution according to the Shapiro–Wilk test, while the mean values were compared by Duncan's multiple range test at $p = 0.05$. Statistical analysis was processed with JMP v. 16.1 (SAS Institute Inc.). The results are presented as mean values and standard deviations (SD).

3. Results and Discussion

The results of growth parameters are presented in Table 1. Significant statistical differences regarding the morphological traits in relation to the different fertilization regimes were recorded. The highest number of leaves/plant was achieved for the SH4 (300:100:100; 19.46) fertilization regime, being significantly different from the rest of the treatments, whereas SH1 (100:100:100) and SH6 (300:300:300) treatments recorded the lowest mean values. The SH3 (200:200:200) and SH5 (300:200:200) treatments recorded the highest fresh weight of leaves/plant (116.41 g and 113.58 g, respectively), whereas plants treated with SH6 and the control treatment had the significantly lowest fresh weight. On the other hand, the highest dry matter content was observed for the control treatment, while the leaves obtained from plants treated with SH4 and SH5 had the lowest overall dry matter content. Our results indicate that apart from the increased amounts of nitrogen, the application of nutrient solution with balanced composition in P and K had a beneficial effect on fresh biomass yield, whereas excessive amount of macronutrients resulted in fresh yields similar to the untreated plants due to reduction in leaves number. These findings corroborate the aspect of minimum nutrients requirements for wild edible species, especially for the case of P and K, which were also confirmed by Polyzos et al. [39] in *Cichorium spinosum* cultivation. Moreover, Papadimitriou et al. [60] suggested that a high ratio of N:K (e.g., 2.38 mol/mol) in nutrient solution resulted in increased yields of leaves and roots in hydroponically grown *Scolymus hispanicus* compared to lower ratios (e.g., 1.59 mol/mol). This contradiction could be attributed to differences in the cropping systems (open hydroponic system vs. pot cultivation), since the availability and uptake of N are not comparable. Regarding the dry matter content, Polyzos et al. [39] also reported increased values of dry matter in untreated plants *C. spinosum* plants, suggesting stressful conditions due to nutrients deprivation, a finding that is in agreement with the results of our study.

Table 1. The effect of the type of substrate on the weight of plant (g), the weight of leaves/plant (g), the number of leaves/plant and the dry matter of leaves (%) of *S. hispanicus* plants.

Treatments	Number of Leaves/Plant	Weight of Leaves/Plant (g)	Dry Matter of Leaves (%)
SHC	15.57 ± 2.28 (c)	94.24 ± 8.71 (c)	11.67 ± 1.51 (a)
SH1	15.00 ± 2.45 (d)	102.27 ± 11.24 (b)	9.73 ± 2.70 (d)
SH2	16.17 ± 1.59 (bc)	99.49 ± 8.64 (b)	10.52 ± 1.13 (b)
SH3	15.75 ± 1.29 (c)	116.41 ± 7.28 (a)	10.16 ± 2.32 (c)
SH4	19.46 ± 1.14 (a)	99.36 ± 4.52 (b)	8.81 ± 0.57 (e)
SH5	16.83 ± 2.25 (b)	113.58 ± 8.09 (a)	8.69 ± 0.25 (e)
SH6	15.11 ± 0.78 (d)	90.67 ± 10.73 (c)	10.02 ± 1.50 (c)

Mean values in the same column followed by different letters are significantly different at $p < 0.05$ according to Duncan's multiple range test.

Table 2 presents the results of chlorophyll content (SPAD index, leaf area and specific leaf area). SPAD index values were the highest when plants were treated with SH4 and SH1 treatments. Similarly, the plants that received SH4 treatment formed the highest leaf area (1721.63 cm^2), without being significantly different from SH1 treatment, while all the tested fertilization regimes had higher SAPD index and leaf area values than the untreated plants (SHC treatment). This finding is in line with the highest number of leaves per plant, which was recorded for the SH4 treatment, thus indicating that the leaf area increased due to the formation of more leaves instead of the development of bigger leaves. Similarly, control treatment recorded the lowest overall value of specific leaf area (12.65 m^2/kg), while the significantly highest values were measured for plants treated with SH5 and

SH4 treatments (16.82 and 16.42 m^2/kg, respectively). The significant effect of fertilization regime on SPAD index value of wild edible leafy greens has been also reported by Polyzos et al. [39] and Fidimundy et al. [46], who studied the growth parameters of *Cichorium spinosum* and *Sanguisorba minor* plants, respectively. On the other hand, Tzortzakis and Klados [64] and Papadimitriou et al. [59] did not record any differences in *C. spinosum* and *S. hispanicus* plants treated with nutrient solutions of different salinity levels. According to Di Mola et al. [65], SPAD index values of baby spinach and lamb's lettuce leaves were positively correlated with nitrogen availability, while Karkanis et al. [66] highlighted the importance of harvesting stage on this parameter. Moreover, Fontana et al. [55] suggested that apart from total nitrogen availability, the nitrogen form may also affect chlorophyll content in purslane plants cultivated in a soilless hydroponic system. Based on our results, the increased amounts of nitrogen combined with low amounts of P and K (SH4) or a balanced solution of N:P:K (SH1) were beneficial to chlorophyll content and leaf area, whereas Polyzos et al. [39] suggested that the untreated plants or those that received a nutrient solution that contained 200:200:200 ppm of N, P and K had the highest overall SPAD values. Therefore, it could be assumed that each species may respond differently to fertilization regime, while in our case, the excessive inputs of micronutrients in *S. hispanicus* (e.g., SH6) were not positively correlated with visual quality and fresh yield of leaves.

Table 2. The effect of the fertilization regime on chlorophyll content of the leaves (SPAD values), leaf area (cm^2) and specific leaf area (m^2/kg) of *S. hispanicus* plants.

Treatments	Chlorophyll Content	Leaf Area (cm^2)	Specific Leaf Area (m^2/kg)
SHC	32.14 ± 1.15 (f)	1298.02 ± 179.51 (e)	12.65 ± 1.18 (d)
SH1	37.26 ± 1.77 (a)	1715.79 ± 153.24 (a)	14.85 ± 1.17 (b)
SH2	34.83 ± 1.39 (e)	1402.48 ± 171.05 (c)	13.72 ± 1.29 (c)
SH3	35.72 ± 1.59 (c)	1579.85 ± 202.66 (b)	13.45 ± 1.09 (c)
SH4	37.58 ± 1.25 (a)	1721.63 ± 152.60 (a)	16.42 ± 1.21 (a)
SH5	36.19 ± 2.15 (b)	1614.04 ± 140.34 (b)	16.82 ± 0.29 (a)
SH6	35.35 ± 1.31 (d)	1381.79 ± 169.87 (d)	15.18 ± 1.52 (b)

Mean values in the same column followed by different letters are significantly different at $p < 0.05$ according to Duncan's multiple range test.

The nutritional profile of *S. hispanicus* leaves in relation to fertigation regime is presented in Table 3, where a variable response was detected. Total fat, crude protein and ash content were negatively affected by fertigation, since the highest overall values were recorded for the untreated plants. In contrast, dietary fiber and carbohydrates content were significantly higher for plants treated with SH3 and SH5 treatments, resulting in a higher energy content for the latter treatment. These results are within the range of the values reported by García-Herrera et al. [56] who evaluated the proximate composition of wild golden thistle plants collected from different sites and in different years. However, the authors of that study detected a great variation among the tested samples, which indicates a significant effect of the genotype and the growing conditions on the nutritional value of the species. Wild *S. hispanicus* plants are considered a rich source of dietary fiber [3,56] and total carbohydrates [23], which in the case of our study, its content was significantly increased by fertigation regimes, whereas fat and ash contents were low compared to other wild edible species [23]. According to the literature, commercial cultivation practices may affect the nutritional value of wild species depending on the species [61,67], allowing us to regulate the quality of the final product and improve the content of beneficial compounds. However, this is not always the case, as for example, Disciglio et al. [18] did not observe a significant difference in protein content between wild and cultivated plants of *C. intybus*, *Borago officinalis* and *Diplotaxis tenuifolia*, whereas Polyzos et al. [39] recorded a decrease in protein and ash content when *C. spinosum* plants were treated with nutrient solutions similar to our study.

Table 3. Nutritional profile and energy content of *Scolymus hispanicus* leaves in relation to fertigation regime (Mean ± SD).

	SHC	SH1	SH2	SH3	SH4	SH5	SH6
				Nutritional profile (g/100 g dw)			
Total fat	2.9 ± 0.1 [a]	2.8 ± 0.1 [a]	2.3 ± 0.1 [c]	2.6 ± 0.1 [b]	2.61 ± 0.02 [b]	2.73 ± 0.02 [a,b]	2.2 ± 0.1 [c]
Crude protein	11.84 ± 0.03 [a]	10.6 ± 0.1 [b]	10.73 ± 0.41 [b]	10.8 ± 0.5 [b]	9.8 ± 0.1 [d]	9.8 ± 0.1 [d]	10.4 ± 0.1 [c]
Ash	11.8 ± 0.4 [a]	11.5 ± 0.4 [b]	9.856 ± 0.003 [d]	11.4 ± 0.08 [b]	10.04 ± 0.09 [d]	9.9 ± 0.1 [d]	11.02 ± 0.18 [c]
Total dietary fiber	37.1 ± 0.1 [e]	37.0 ± 0.2 [e]	40.2 ± 1.1 [b]	40.7 ± 0.2 [a]	38.81 ± 0.04 [d]	36.269 ± 0.002 [f]	39.6 ± 0.6 [c]
Carbohydrates	36.36 ± 0.35 [d]	38.04 ± 0.57 [c]	36.9 ± 1.1 [d]	34.5 ± 0.6 [e]	38.7 ± 0.1 [b]	41.3 ± 0.2 [a]	36.8 ± 0.6 [d]
				Energy (kcal/100 g dw)			
	292.9 ± 1.6 [c]	294.2 ± 1.1 [b]	291.8 ± 1.4 [c]	285.7 ± 0.1 [f]	295.3 ± 0.4 [b]	301.4 ± 0.3 [a]	287.9± 1.0 [e]

Mean values in the same column followed by different letters are significantly different at $p < 0.05$ according to Duncan's multiple range test. dw: dry weight.

Organic acid composition is presented in Table 4. The main detected compounds were oxalic, quinic and malic acid followed by shikimic and citric acid, while traces of fumaric acids were also present in the studied samples. A varied response to fertigation regime was observed, with higher amounts of oxalic, quinic and total organic acids being detected in the untreated plants. Malic acid was the highest in SH treatment, while the SH3 treatment resulted in the highest amounts of shikimic and citric acid. On the other hand, the SH4 treatment resulted in the lowest values for all the detected compounds (except for the case of quinic acid where the lowest content was recorded for SH2 treatment) and consequently in the lowest content of total organic acids. According to the literature, the increased inputs of nitrogen are associated with high amounts of oxalic acid [61,67], while the nitrogen form may affect the accumulation of this particular organic acid or total organic acids [68]. Moreover, Dias et al. [69] reported a decrease in oxalic acid content in cultivated *Achillea millefolium* plants compared to unattended ones. This contradiction could be due to the fact that in our study, the control plants were subjected to stress conditions due to nutrient deprivation and P in particular. According to Le Roux et al. [70], P deficiency is associated with the synthesis and accumulation of organic acids which tend to decrease nitrogen assimilation. Another possible explanation could be related with the ratios and the total amounts of macronutrients applied in the tested fertigation regimes, which may result in synergistic or antagonistic effects that consequently affect nutrient assimilation and impair plant physiology and metabolism [71]. According to Aboyeji et al. [72], high amounts of P may have detrimental effects on the yield of groundnut plants due to antagonism between P and Zn that may affect plant growth and development. Moreover, excessive amounts of nitrogen are associated with reduced uptake of other nutrients, which result in stressful conditions and consequently in the accumulation of organic acids [73,74]. Therefore, based on our results and the literature reports, further studies are needed to reveal uptake and translocation of nutrients from roots to upper parts in order to reveal the mechanisms involved in organic acids biosynthesis as part of the antioxidant and osmoregulatory mechanisms of plants. In any case, the reduction of organic acids and oxalic acid in particular after the application of the tested fertigation regimes can be considered a positive impact on the quality of the final product, since oxalic acid is an antinutritional factor, and high intake (>5 g per day) may have severe health effects [51].

Tocopherol and free sugar compositions are presented in Table 5. Alpha-tocopherol was the only detected isoform of vitamin in all the studied samples, while except for the case of SH5 treatment, the rest of the fertigation regimes resulted in a significant decrease compared to untreated plants. The detected amounts were different from those reported by Petropoulos et al. [21] (0.68 μg/100 g fw) and Marmouzi et al. [75] (0.54 mg/100 g) and in the same range with the study of Morales et al. [24] (0.02 mg/100 g fw) and Vardavas et al. [22] (0.038 mg/100 g fw). Moreover, in contrast to our study, β- and γ-tocopherols were also detected in *S. hispanicus* leaves, which could be due to different growing conditions and genotypic variability [76]. The variable response to fertilization regime was also reported in the study of Polyzos et al. [39] who evaluated the response of *C. spinosum* plants as similar to our study fertigation regimes and recorded the highest α- and total tocopherols content for the treatment of 300:200:200 ppm of N:P:K. It seems that high amounts of nitrogen combined with a balanced content of P and K in the nutrient solution may improve tocopherol composition, a finding which has been confirmed in other species [67,77].

Table 4. Organic acids (mg/100 g dw) content in *S. hispanicus* leaves in relation to fertigation regime (mean ± SD).

	SHC	SH1	SH2	SH3	SH4	SH5	SH6
Oxalic acid	3.13 ± 0.02 ᵃ	2.28 ± 0.01 ᵈ	2.08 ± 0.01 ᵉ	2.92 ± 0.02 ᵇ	1.8 ± 0.01 ᵍ	2.440 ± 0.001 ᶜ	1.980 ± 0.008 ᶠ
Quinic acid	2.86 ± 0.05 ᵃ	2.37 ± 0.02 ᶜ	2.15 ± 0.01 ᵉ	2.8 ± 0.1 ᵇ	2.2 ± 0.1 ᵈ	2.39 ± 0.02 ᶜ	2.2 ± 0.02 ᵈ
Malic acid	2.250 ± 0.002 ᵇ	2.37 ± 0.02 ᵃ	1.660 ± 0.004 ᵉ	1.88 ± 0.02 ᵈ	1.61 ± 0.04 ᶠ	1.95 ± 0.03 ᶜ	1.66 ± 0.01 ᵉ
Shikimic acid	0.026 ± 0.006 ᵇ	0.021 ± 0.001 ᵉ	0.016 ± 0.001 ᵍ	0.031 ± 0.001 ᵃ	0.018 ± 0.001 ᶠ	0.025 ± 0.002 ᶜ	0.0223 ± 0.0002 ᵈ
Citric acid	0.86 ± 0.02 ᵇ	0.84 ± 0.02 ᶜ	0.82 ± 0.03 ᶜ,ᵉ	0.903 ± 0.003 ᵃ	0.77 ± 0.03 ᶠ	0.83 ± 0.01 ᶜ,ᵈ	0.82 ± 0.01 ᵈ,ᵉ
Fumaric acid	tr	tr	tr	tr	tr	tr	tr
Sum	9.13 ± 0.02 ᵃ	7.88 ± 0.01 ᶜ	6.74 ± 0.06 ᵉ	8.5 ± 0.1 ᵇ	6.4 ± 0.1 ᵍ	7.63 ± 0.01 ᵈ	6.68 ± 0.02 ᶠ

Mean values in the same column followed by different letters are significantly different at $p < 0.05$ according to Duncan's multiple range test. Calibration curves for organic acids: oxalic acid ($y = 8 \times 10^6x + 331789$, $R^2 = 0.9912$); quinic acid ($y = 692575x + 11551$; $R^2 = 0.9983$); citric acid ($y= 968367x–12295$, $R^2= 0.9974$); shikimic acid ($y = 5 \times 10^7x + 567119$, $R^2 = 0.9903$); fumaric acid ($y = 9 \times 10^7x–100894$, $R^2 = 0.9986$); tr: traces; dw: dry weight.

Table 5. Tocopherol (mg/100 g dw) and free sugar (g/100 g dw) content in *S. hispanicus* leaves in relation to fertigation regime (mean ± SD).

	SHC	SH1	SH2	SH3	SH4	SH5	SH6
				Tocopherols (mg/100 g dw)			
α-Tocopherol	0.095 ± 0.007 ᵇ	0.070 ± 0.003 ᵈ	0.050 ± 0.002 ᵉ	0.080 ± 0.003 ᶜ	0.021 ± 0.001 ᶠ	0.25 ± 0.01 ᵃ	0.02 ± 0.001 ᶠ
				Free sugars (g/100 g dw)			
Fructose	4.2 ± 0.2 ᵇ	4.1 ± 0.1 ᵇ	3.9 ± 0.1 ᶜ	4.75 ± 0.04 ᵃ	4.72 ± 0.01 ᵃ	3.86 ± 0.04 ᶜ	3.6 ± 0.1 ᵈ
Glucose	6.6 ± 0.2 ᵃ	5.9 ± 0.1 ᶜ	6.3 ± 0.2 ᵇ	5.9 ± 0.1 ᶜ	6.2 ± 0.2 ᵇ	5.4 ± 0.1 ᵉ	5.48 ± 0.04 ᵈ
Sucrose	1.8 ± 0.02 ᶠ	1.94 ± 0.01 ᵉ	1.77 ± 0.02 ᶠ	2.1 ± 0.1 ᵈ	2.41 ± 0.02 ᶜ	2.5 ± 0.1 ᵇ	2.8 ± 0.2 ᵃ
Trehalose	1.21 ± 0.03 ᵇ	1.34 ± 0.03 ᵃ	1.19 ± 0.03 ᵇ	0.88 ± 0.03 ᶜ	0.66 ± 0.02 ᵈ	0.89 ± 0.01 ᶜ	1.3 ± 0.2 ᵃ
Sum	13.8 ± 0.4 ᵇ	13.35 ± 0.04 ᶜ	13.1 ± 0.2 ᵈ	13.68 ± 0.28 ᵇ	13.97 ± 0.19 ᵃ	12.65 ± 0.01 ᵉ	13.18 ± 0.002 ᵈ

Mean values in the same column followed by different letters are significantly different at $p < 0.05$ according to Duncan's multiple range test; dw: dry weight.

Regarding the free sugar composition, glucose was the main detected free sugar, followed by fructose, sucrose and trehalose. A variable response was recorded in relation to the tested fertigation regimes. In particular, the highest glucose content was detected in control plants, while fructose content was the highest in SH3 and SH4 treatments. Similarly, sucrose and trehalose were significantly higher for the SH6 and SH1 and SH6 treatments, respectively. Finally, the highest and lowest total free sugar content was recorded for SH4 and SH% treatments, respectively. To the best of our knowledge, this is the first report regarding the free sugar composition and no references are available for comparison purposes. However, Polyzos et al. [39] who studied the same fertigation regimes in *C. spinosum* plants also reported a varied response, while similar fluctuations have been suggested in other crops due to growing conditions and nutrients availability [78,79]. Regulation of free sugar composition through the application of tailor-made fertigation regimes could be a cost-effective means to improve the quality of the final products of wild edible species, since the increased sugar content could be associated with improved taste and organoleptic properties.

Mineral composition of *S. hispanicus* leaves is presented in Table 6. A varied response was recorded without specific trends in the effects of the tested fertigation regimes being observed. In particular, the untreated plants had the highest content of K and Zn, while plants treated with SH4 treatment recorded the highest content of Na, Ca and Mg. Moreover, SH1 and SH2 treatments had significantly higher Fe content compared to the rest of the treatments, while Mn and Cu content was significantly higher for SH1 treatment. The range of minerals detected in our study was in the same range as the values reported by García-Herrera et al. [56] with slight variations in the case of Ca and Cu, while Papadimitriou et al. [59] reported higher values for Na, K, Ca and Mg. However, it has to be noted that García-Herrera et al. [56] recorded a high variability in minerals profile depending on the collection site and year, and they suggested a significant impact on growing conditions, while they determined the mineral content of the midribs *S. hispanicus* instead of whole leaves, which were evaluated in our study. According to Rietra et al. [72], significant interaction may occur among plant macro- and micronutrients, which may negatively or positively affect plant growth and crop yield. This is evident in our study where the varied amounts of N:P:K applied through fertigation resulted in a varied response regarding the mineral composition of *S. hispanicus* leaves. According to Fageria and Oliveira [80], P is the most influential nutrient since its imbalance may severely affect crop yield, a finding that is in agreement with our study where the highest overall fresh yield was recorded for the treatments where 200 ppm of P was applied (SH3 and SH5). Moreover, other studies report significant antagonistic effects between P and Mg or Ca [81,82], which coincide with the findings of our study where the highest Mg and Ca contents were recorded in SH4 treatment where 300:100:100 ppm of N:P:K was applied. Therefore, it could be suggested that the regulation of fertilization regime may favor crop yield and improve the mineral profile of *S. hispanicus* leaves at the same time. However, the impact of growing conditions and genotype should be further investigated.

Table 6. Mineral composition in *S. hispanicus* leaves in relation to fertigation regime (mean ± SD).

	SHC	SH1	SH2	SH3	SH4	SH5	SH6
[K]/(g/Kg)	27.1 ± 0.7 ^{a}	20.4 ± 0.7 ^{b}	13.4 ± 0.4 ^{e}	19 ± 1 ^{c}	12.70 ± 0.02 ^{f}	13.98 ± 0.17 ^{d}	18.9 ± 0.7 ^{c}
[Na]/(mg/Kg)	5757 ± 182 ^{d}	5713 ± 228 ^{d}	6139 ± 75 ^{c}	6174 ± 166 ^{c}	7396 ± 110 ^{a}	6717 ± 103 ^{b}	6768 ± 302 ^{b}
[Ca]/(g/Kg)	8.02 ± 0.06 ^{c}	8.5 ± 0.4 ^{b}	8.2 ± 0.1 ^{b,c}	8.3 ± 0.4 ^{b}	9.6 ± 0.4 ^{a}	7.2 ± 0.3 ^{e}	7.70 ± 0.02 ^{d}
[Mg]/(g/Kg)	1.98 ± 0.04 ^{f}	2.30 ± 0.03 ^{e}	2.6 ± 0.1 ^{b}	2.4 ± 0.1 ^{c,d}	3.6 ± 0.1 ^{a}	2.44 ± 0.01 ^{c}	2.36 ± 0.02 ^{d,e}
[Fe]/(mg/Kg)	172 ± 6 ^{e}	222 ± 2 ^{a,b}	223 ± 8 ^{a}	218 ± 7 ^{b}	194 ± 8 ^{c}	185 ± 3 ^{d}	184 ± 7 ^{d}
[Mn]/(mg/Kg)	116.8 ± 0.2 ^{f}	155 ± 9 ^{a}	142 ± 3 ^{c}	123 ± 5 ^{e}	135 ± 2 ^{d}	122 ± 6 ^{e}	148 ± 2 ^{b}
[Cu]/(mg/Kg)	4.3 ± 0.1 ^{g}	5.7 ± 0.2 ^{a}	5.2 ± 0.2 ^{c,e}	5.3 ± 0.2 ^{b,c}	5.4 ± 0.2 ^{b}	4.71 ± 0.02 ^{f}	5.2 ± 0.1 ^{e}
[Zn]/(mg/Kg)	48.1 ± 0.8 ^{a}	33.9 ± 0.7 ^{c}	28 ± 1 ^{d}	35.5 ± 0.5 ^{b}	20.5 ± 0.6 ^{f}	20.8 ± 0.3 ^{f}	25.8 ± 0.4 ^{e}

Mean values in the same column followed by different letters are significantly different at $p < 0.05$ according to Duncan's multiple range test.

The fatty acid composition is presented in Table 7. The most abundant compounds were α-linolenic acid (47.8–59.4%), palmitic acid (19.4–24.5%) and linoleic acid (10.44–12.72%), while the polyunsaturated fatty acids (PUFA) were the major class of fatty acids (60.3–70.8%). A varied response was recorded in relation to the fertigation regime without specific trends being observed among the tested treatments. The highest content of the major fatty acids was recorded either for untreated plants (e.g., α-linolenic acid) or for the treatments with low nutrient amounts (e.g., SH1 and SH2 in the case of linoleic and palmitic acid, respectively). The recorded composition was in the same range as the fatty acids profile reported by Morales et al. [83] who detected high amounts of PUFA (57.66%), followed by monounsaturated and saturated fatty acids (MUFA: 34.16% and SFA: 8.19%) in peeled basal leaves. However, the content of individual compounds varied compared to our study, with α-linolenic acid being the most abundant one (30.55%), followed by linoleic and palmitic acids (26.44 and 20.65%, respectively). In contrast to our study, Vardavas et al. [25] reported a different composition of the main fatty acids groups, suggesting MUFA (54.8%) as the most abundant class, followed by SFA (33.7%) and PUFA (11.4%). Moreover, they recorded a balanced ratio of n6/n3 (1.06) and different amounts of individual fatty acids (linoleic acid 33.8%; α-linolenic acid 32% and palmitic acid 30.3%). The variable-reported results could be associated with the tested raw material (whole leaves in our study were compared to peeled basal leaves) as well as to growing conditions and agronomic practices. Considering that the plants in our study were grown in a growth substrate under different fertigation regimes, this could also be a possible explanation for the observed differences as already confirmed by the literature reports [39,45].

Table 7. Fatty acid composition (relative percentage %) in *S. hispanicus* leaves in relation to fertigation regime (mean ± SD).

	SHC	SH1	SH2	SH3	SH4	SH5	SH6
C8:0	nd	0.496 ± 0.018^{e}	0.523 ± 0.001^{d}	0.62 ± 0.03^{c}	0.84 ± 0.03^{a}	0.66 ± 0.02^{b}	nd
C13:0	0.93 ± 0.01^{e}	0.43 ± 0.01^{f}	1.23 ± 0.05^{b}	1.15 ± 0.01^{c}	1.33 ± 0.04^{a}	1.14 ± 0.04^{c}	1.11 ± 0.01^{d}
C14:0	1.256 ± 0.003^{e}	0.83 ± 0.03^{f}	2.5 ± 0.08^{a}	1.63 ± 0.01^{d}	1.86 ± 0.05^{b}	1.76 ± 0.06^{c}	1.87 ± 0.07^{b}
C14:1	0.48 ± 0.01^{e}	0.21 ± 0.01^{f}	0.62 ± 0.01^{c}	0.58 ± 0.03^{d}	0.72 ± 0.02^{a}	0.616 ± 0.005^{c}	0.69 ± 0.03^{b}
C15:0	0.263 ± 0.001^{a}	0.17 ± 0.01^{d}	0.14 ± 0.01^{g}	0.2 ± 0.01^{c}	0.238 ± 0.003^{b}	0.157 ± 0.003^{f}	0.162 ± 0.003^{e}
C16:0	19.4 ± 0.1^{f}	24.34 ± 0.03^{b}	24.5 ± 0.2^{a}	23.1 ± 0.6^{c}	23.3 ± 0.2^{c}	19.9 ± 0.2^{e}	20.4 ± 0.1^{d}
C16:1	$1.76 \pm 0.01^{b,c}$	2.24 ± 0.01^{a}	1.67 ± 0.02^{d}	1.74 ± 0.01^{c}	1.77 ± 0.08^{b}	1.607 ± 0.004^{e}	1.487 ± 0.003^{f}
C17:0	0.54 ± 0.01^{d}	1.251 ± 0.001^{b}	1.16 ± 0.04^{c}	1.25 ± 0.04^{b}	1.58 ± 0.06^{a}	0.39 ± 0.01^{f}	0.448 ± 0.004^{e}
C18:0	2.02 ± 0.07^{f}	2.4 ± 0.1^{d}	3.9 ± 0.1^{a}	3.1 ± 0.1^{b}	3.1 ± 0.1^{b}	2.25 ± 0.03^{e}	2.5 ± 0.1^{c}
C18:1n9c	1.373 ± 0.001^{c}	1.31 ± 0.04^{d}	2.07 ± 0.07^{a}	$1.4 \pm 0.06^{b,c}$	1.44 ± 0.03^{b}	1.2 ± 0.02^{e}	1.4 ± 0.1^{c}
C18:2n6c	11.04 ± 0.35^{c}	12.72 ± 0.03^{a}	10.44 ± 0.05^{e}	11.13 ± 0.09^{c}	10.77 ± 0.09^{d}	11.81 ± 0.04^{b}	11.9 ± 0.7^{b}
C18:3n3	59.4 ± 0.4^{a}	51.9 ± 0.1^{d}	47.8 ± 0.1^{g}	49.9 ± 0.3^{e}	48.9 ± 0.4^{f}	56.3 ± 0.2^{b}	54.97 ± 1.14^{c}
C22:0	0.62 ± 0.02^{e}	0.67 ± 0.01^{e}	1.32 ± 0.06^{c}	1.77 ± 0.05^{a}	1.52 ± 0.01^{b}	0.84 ± 0.01^{d}	1.23 ± 0.36^{c}
C23:0	0.36 ± 0.01^{g}	0.39 ± 0.01^{f}	0.85 ± 0.03^{c}	0.9 ± 0.04^{b}	1.01 ± 0.01^{a}	0.61 ± 0.02^{e}	0.76 ± 0.03^{d}
C24:0	0.617 ± 0.002^{f}	0.68 ± 0.01^{e}	1.271 ± 0.004^{b}	1.58 ± 0.06^{a}	1.58 ± 0.03^{a}	0.82 ± 0.03^{d}	1.06 ± 0.01^{c}
SFA	25.7 ± 0.3^{g}	31.5 ± 0.2^{d}	37.3 ± 0.5^{a}	35.1 ± 1^{c}	36.1 ± 0.5^{b}	28.3 ± 0.4^{f}	29.4 ± 0.7^{e}
MUFA	2.498 ± 0.025^{c}	2.63 ± 0.02^{b}	2.43 ± 0.03^{d}	2.52 ± 0.05^{c}	2.73 ± 0.1^{a}	2.38 ± 0.01^{e}	2.34 ± 0.04^{f}
PUFA	71.8 ± 0.7^{a}	65.9 ± 0.1^{d}	60.3 ± 0.2^{g}	62.4 ± 0.4^{e}	61.1 ± 0.5^{f}	69.3 ± 0.2^{b}	68 ± 2^{c}
n6/n3	0.186 ± 0.005^{f}	0.245 ± 0.001^{a}	0.218 ± 0.001^{d}	0.223 ± 0.001^{b}	0.220 ± 0.001^{c}	0.209 ± 0.001^{e}	0.216 ± 0.008^{d}
PUFA/SFA	2.794 ± 0.005^{a}	2.09 ± 0.01^{d}	1.61 ± 0.02^{g}	1.76 ± 0.05^{e}	1.69 ± 0.02^{f}	2.44 ± 0.03^{b}	2.32 ± 0.02^{c}

Mean values in the same column followed by different letters are significantly different at $p < 0.05$ according to Duncan's multiple range test. Fatty acids are expressed as relative percentage of each fatty acid. C8:0—caprylic acid; C13:0—tridecanoic acid; C14:0—myristic acid; C14:1—tetradecanoic acid; C15:0—pentadecanoic acid; C15:1; C16:0—palmitic acid; C16:1—palmitoleic acid; C17:0—heptadecanoic acid; C18:0—stearic acid; C18:1n9c—oleic acid; C18:2n6c—linoleic acid; C18:3n3—linolenic acid; C22:0—Behenic acid; C23:0—tricosanoic acid; C24:0—lignoceric acid. SFA—saturated fatty acids; MUFA—monounsaturated fatty acids; PUFA—polyunsaturated fatty acids; nd—not detected.

4. Conclusions

The domestication of wild edible species is pivotal for the introduction and integration of such alternative crops in the existing farming systems, especially in small-scale farms of the Mediterranean. Fertilization is one of the most effective and common agronomic practices that is applied in commercial farming and significantly contributes to yield increase in various crops. Considering the lack of information regarding the best practice guides and fertilization regimes for *S. hispanicus*, this study aimed to evaluate how different amounts of nutrients can affect yield and quality of edible leaves. In particular, treatments with moderate amounts of P and K (SH3 and SH5) recorded the highest fresh yield, whereas the highest inputs (SH6) negatively affected fresh biomass yield. The highest leaf area was recorded for SH4 treatment and coincided with the highest numbers of leaves. Apart from the highest fresh weight, SH3 and SH5 treatments were the most beneficial for dietary fiber and carbohydrates content, which are important nutritional parameters. On the other hand, SH4 treatment recorded the lowest oxalic acid content, which is considered an anti-nutritional factor, while it was beneficial to mineral profile (Ca and Mg content), fructose and total sugars content. Finally, treatments SHC, SH1 and SH2 were beneficial to the major fatty acids content (α-linolenic, linoleic and palmitic acid, respectively). In conclusion, it could be suggested that the regulation of nutrient solution seems to be an effective practice to increase fresh yield in *S. hispanicus* with low to moderate inputs, without compromising the nutritional profile of the edible product. This is an important finding for introducing sustainable practices in the existing farming systems, since the commercial cultivation of wild species such as *S. hispanicus* reduces the risk of genetic erosion due to irrational harvesting and increases the biodiversity of agroecosystems and its resilience to climate change conditions.

Author Contributions: Conceptualization, Y.R., L.B. and S.A.P.; methodology, N.P., B.H.P., M.C. and M.I.D.; formal analysis, B.H.P. and M.I.D.; investigation, N.P., M.C., B.H.P. and M.I.D.; resources, L.B. and S.A.P.; data curation, N.P., B.H.P. and M.I.D.; writing—original draft preparation, N.P., B.H.P. and M.I.D.; writing—review and editing, A.A., Y.R., L.B. and S.A.P.; visualization, S.A.P.; supervision, L.B and S.A.P.; project administration, L.B. and S.A.P.; funding acquisition, S.A.P. All authors have read and agreed to the published version of the manuscript.

Funding: This work was funded by the General Secretariat for Research and Technology of Greece and PRIMA foundation under the project VALUEFARM (PRIMA2019-11). The authors are also grateful to the Foundation for Science and Technology (FCT, Portugal) for financial support through national funds FCT/MCTES (PIDDAC) to CIMO (UIDB/00690/2020 and UIDP/00690/2020) and SusTEC (LA/P/0007/2021); for the grant of B.H. Paschoalinotto and for the financial support within the scope of the Project PRIMA Section 2—Multi-topic 2019: VALUEFARM (PRIMA/0009/2019); and to L. Barros, and M.I. Dias, FCT, P.I., through an institutional scientific employment program contract for their contracts (CEEC Institutional).

Institutional Review Board Statement: Not applicable.

Informed Consent Statement: Not applicable.

Data Availability Statement: Data are available upon request.

Acknowledgments: Not applicable.

Conflicts of Interest: The authors declare no conflict of interest.

References

1. Aguiar, S.; Texeira, M.; Garibaldi, L.A.; Jobbágy, E.G. Global changes in crop diversity: Trade rather than production enriches supply. *Glob. Food Sec.* **2020**, *26*, 100385. [CrossRef]
2. Balié, J. The trade-offs of healthy food from sustainable agriculture in the Global South. *Glob. Food Sec.* **2020**, *26*, 100384. [CrossRef]
3. De Cortes Sánchez-Mata, M.; Tardío, J.; Sánchez-Mata, M.; Tardío, J. *Mediterranean Wild Edible Plants: Ethnobotany and Food Composition Tables*; Sánchez-Mata, M.d.C., Tardío, J., Eds.; Springer: New York, NY, USA, 2016; ISBN 9789048128525.
4. Corrêa, R.C.G.; Di Gioia, F.; Ferreira, I.C.F.R.; Petropoulos, S.A. Wild greens used in the Mediterranean diet. In *The Mediterranean Diet: An Evidence-Based Approach*; Preedy, V., Watson, R., Eds.; Academic Press: London, UK, 2020; pp. 209–228. ISBN 9788578110796.
5. Chatzopoulou, E.; Carocho, M.; Di Gioia, F.; Petropoulos, S.A. The beneficial health effects of vegetables and wild edible greens: The case of the mediterranean diet and its sustainability. *Appl. Sci.* **2020**, *10*, 9144. [CrossRef]
6. Bacchetta, L.; Visioli, F.; Cappelli, G.; Caruso, E.; Martin, G.; Nemeth, E.; Bacchetta, G.; Bedini, G.; Wezel, A.; van Asseldonk, T.; et al. A manifesto for the valorization of wild edible plants. *J. Ethnopharmacol.* **2016**, *191*, 180–187. [CrossRef] [PubMed]
7. Luczaj, L.; Pieroni, A.; Tardío, J.; Pardo-De-Santayana, M.; Sõukand, R.; Svanberg, I.; Kalle, R.; Łuczaj, Ł.; Pieroni, A.; Tardío, J.; et al. Wild food plant use in 21st century Europe: The disappearance of old traditions and the search for new cuisines involving wild edibles. *Acta Soc. Bot. Pol.* **2012**, *81*, 359–370. [CrossRef]
8. Laghetti, G.; Bisignano, V.; Urbano, M. Genetic resources of vegetable crops and their safeguarding in Italy. *Hortic. Int. J.* **2018**, *2*, 72–74. [CrossRef]
9. Turner, N.J.; Luczaj, L.J.; Migliorini, P.; Pieroni, A.; Dreon, A.L.; Sacchetti, L.E.; Paoletti, M.G. Edible and tended wild plants, traditional ecological knowledge and Agroecology. *CRC Crit. Rev. Plant Sci.* **2011**, *30*, 198–225. [CrossRef]
10. Nebel, S.; Pieroni, A.; Heinrich, M. Ta chórta: Wild edible greens used in the Graecanic area in Calabria, Southern Italy. *Appetite* **2006**, *47*, 333–342. [CrossRef]
11. Polo, S.; Tardío, J.; Vélez-del-Burgo, A.; Molina, M.; Pardo-de-Santayana, M. Knowledge, use and ecology of golden thistle (*Scolymus hispanicus* L.) in Central Spain. *J. Ethnobiol. Ethnomed.* **2009**, *5*, 42. [CrossRef]
12. Tanji, A.; Nassif, F. Edible weeds in Morocco. *Weed Technol.* **1995**, *9*, 617–620. [CrossRef]
13. Hadjichambis, A.C.; Paraskeva-Hadjichambi, D.; Della, A.; Elena Giusti, M.; De Pasquale, C.; Lenzarini, C.; Censorii, E.; Reyes Gonzales-Tejero, M.; Patricia Sanchez-Rojas, C.; Ramiro-Gutierrez, J.M.; et al. Wild and semi-domesticated food plant consumption in seven circum-Mediterranean areas. *Int. J. Food Sci. Nutr.* **2008**, *59*, 383–414. [CrossRef]
14. Leonti, M.; Nebel, S.; Rivera, D.; Heinrich, M. Wild gathered food plants in the European Mediterranean: A comparative analysis. *Econ. Bot.* **2006**, *60*, 130–142. [CrossRef]
15. Lentini, F.; Venza, F. Wild food plants of popular use in Sicily. *J. Ethnobiol. Ethnomed.* **2007**, *3*, 15. [CrossRef]
16. Aceituno-Mata, L.; Tardío, J.; Pardo-de-Santayana, M. The persistence of flavor: Past and present Uuse of wild food plants in Sierra Norte de Madrid, Spain. *Front. Sustain. Food Syst.* **2021**, *4*, 610238. [CrossRef]
17. Ceccanti, C.; Landi, M.; Benvenuti, S.; Pardossi, A.; Guidi, L. Mediterranean wild edible plants: Weeds or "new functional crops"? *Molecules* **2018**, *23*, 2299. [CrossRef]

18. Disciglio, G.; Tarantino, A.; Frabboni, L.; Gagliardi, A.; Michela, M.; Tarantino, E.; Gatta, G.; Beta, L.; Miller, F.; Cichorium, L.; et al. Qualitative characterisation of cultivated and wild edible plants: Mineral elements, phenols content and antioxidant capacity. *Ital. J. Agron.* **2017**, *12*, 383–394. [CrossRef]
19. Rubio, B.; Villaescusa, L.; Diaz, A.M.; Fernandez, L.; Martin, T. Flavonol Glycosides from *Scolymus hispanicus* and *Jasonia glutinosa*. *Planta Med.* **1995**, *61*, 583. [CrossRef]
20. Rubio, B.; Diaz, A.M.; Velazquez, M.P.; Villaescusa, L. Caffeoyl and flavonoid compounds in *Scolymus hispanicus*. *Planta Med.* **1991**, *57*, A130. [CrossRef]
21. Petropoulos, S.A.; Fernandes, Â.; Tzortzakis, N.; Sokovic, M.; Ciric, A.; Barros, L.; Ferreira, I.C.F.R. Bioactive compounds content and antimicrobial activities of wild edible Asteraceae species of the Mediterranean flora under commercial cultivation conditions. *Food Res. Int.* **2019**, *119*, 859–868. [CrossRef]
22. Vardavas, C.I.; Majchrzak, D.; Wagner, K.H.; Elmadfa, I.; Kafatos, A. The antioxidant and phylloquinone content of wildly grown greens in Crete. *Food Chem.* **2006**, *99*, 813–821. [CrossRef]
23. Tbatou, M.; Kabil, M.; Belahyan, A.; Belahsen, R. Dietary potential of some forgotten wild leafy vegetables from Morocco. *Int. Food Res. J.* **2018**, *25*, 1829–1836.
24. Morales, P.; Ferreira, I.C.F.R.; Carvalho, A.M.; Sánchez-Mata, M.C.; Cámara, M.; Fernández-Ruiz, V.; Pardo-de-Santayana, M.; Tardío, J. Mediterranean non-cultivated vegetables as dietary sources of compounds with antioxidant and biological activity. *LWT Food Sci. Technol.* **2014**, *55*, 389–396. [CrossRef]
25. Vardavas, C.I.; Majchrzak, D.; Wagner, K.H.; Elmadfa, I.; Kafatos, A. Lipid concentrations of wild edible greens in Crete. *Food Chem.* **2006**, *99*, 822–834. [CrossRef]
26. Gray, D.J.; Trigiano, R.N. Towards a more sustainable agriculture. *CRC Crit. Rev. Plant Sci.* **2011**, *30*, 1. [CrossRef]
27. Correa, R.C.G.; Di Gioia, F.; Ferreira, I.; SA, P. Halophytes for future horticulture: The case of small-scale farming in the Mediterranean basin. In *Halophytes for Future Horticulture: From Molecules to Ecosystems towards Biosaline Agriculture*; Marius-Nicusor, G., Ed.; Springer Nature: Cham, Switzerland, 2020; pp. 1–28. ISBN 9783030178543.
28. Panta, S.; Flowers, T.; Lane, P.; Doyle, R.; Haros, G.; Shabala, S. Halophyte agriculture: Success stories. *Environ. Exp. Bot.* **2014**, *107*, 71–83. [CrossRef]
29. Karkanis, A.; Polyzos, N.; Kompocholi, M.; Petropoulos, S.A. Rock samphire, a candidate crop for saline agriculture: Cropping practices, chemical composition and health effects. *Appl. Sci.* **2022**, *12*, 737. [CrossRef]
30. Pereira, A.G.; Fraga-Corral, M.; Garciá-Oliveira, P.; Jimenez-Lopez, C.; Lourenço-Lopes, C.; Carpena, M.; Otero, P.; Gullón, P.; Prieto, M.A.; Simal-Gandara, J. Culinary and nutritional value of edible wild plants from northern Spain rich in phenolic compounds with potential health benefits. *Food Funct.* **2020**, *11*, 8493–8515. [CrossRef]
31. Mikropoulou, E.V.; Vougogiannopoulou, K.; Kalpoutzakis, E.; Sklirou, A.D.; Skaperda, Z.; Houriet, J.; Wolfender, J.-L.; Trougakos, I.P.; Kouretas, D.; Halabalaki, M.; et al. Phytochemical composition of the decoctions of Greek edible greens (chórta) and evaluation of antioxidant and cytotoxic properties. *Molecules* **2018**, *23*, 1541. [CrossRef]
32. Simopoulos, A.P. Omega-3 fatty acids and antioxidants in edible wild plants. *Biol. Res.* **2004**, *37*, 263–277. [CrossRef]
33. Psaroudaki, A.; Dimitropoulakis, P.; Constantinidis, T.; Katsiotis, A.; Skaracis, G.N. Ten indigenous edible plants: Contemporary use in eastern Crete, Greece. *Cult. Agric. Food Environ.* **2012**, *34*, 172–177. [CrossRef]
34. Dogan, Y. Traditionally used wild edible greens in the Aegean Region of Turkey. *Acta Soc. Bot. Pol.* **2012**, *81*, 329–342. [CrossRef]
35. Trichopoulou, A.; Vasilopoulou, E.; Hollman, P.; Chamalides, C.; Foufa, E.; Kaloudis, T.; Kromhout, D.; Miskaki, P.; Petrochilou, I.; Poulima, E.; et al. Nutritional composition and flavonoid content of edible wild greens and green pies: A potential rich source of antioxidant nutrients in the Mediterranean diet. *Food Chem.* **2000**, *70*, 319–323. [CrossRef]
36. Pinela, J.; Carvalho, A.M.; Ferreira, I.C.F.R. Wild edible plants: Nutritional and toxicological characteristics, retrieval strategies and importance for today 's society. *Food Chem. Toxicol.* **2017**, *110*, 165–188. [CrossRef] [PubMed]
37. Nebel, S.; Heinrich, M. Ta chórta: A comparative ethnobotanical-linguistic study of wild food plants in a graecanic area in Calabria, Southern Italy. *Econ. Bot.* **2009**, *63*, 78–92. [CrossRef]
38. Molina, M.; Tardío, J.; Aceituno-Mata, L.; Morales, R.; Reyes-García, V.; Pardo-de-Santayana, M. Weeds and food diversity: Natural yield assessment and future alternatives for traditionally consumed wild vegetables. *J. Ethnobiol.* **2014**, *34*, 44–67. [CrossRef]
39. Polyzos, N.; Paschoalinotto, B.H.; Compocholi, M.; Pinela, J.; Heleno, S.A.; Calhelha, R.C.; Dias, M.I.; Barros, L.; Petropoulos, S.A. Fertilization of pot-grown *Cichorium spinosum* L.: How it can affect plant growth, chemical profile, and bioactivities of edible parts? *Horticulturae* **2022**, *8*, 890. [CrossRef]
40. Papafilippaki, A.; Nikolaidis, N.P. Comparative study of wild and cultivated populations of *Cichorium spinosum*: The influence of soil and organic matter addition. *Sci. Hortic.* **2020**, *261*, 108942. [CrossRef]
41. Chatzigianni, M.; Aliferis, K.A.; Ntatsi, G.; Savvas, D. Effect of N supply level and N source ratio on *Cichorium spinosum* L. metabolism. *Agronomy* **2020**, *10*, 952. [CrossRef]
42. D'Imperio, M.; Parente, A.; Montesano, F.F.; Renna, M.; Logrieco, A.F.; Serio, F. Boron biofortification of *Portulaca oleracea* L. Through soilless cultivation for a new tailored crop. *Agronomy* **2020**, *10*, 999. [CrossRef]
43. Montoya-García, C.O.; Volke-Haller, V.; Trinidad-Santos, A.; Villanueva-Verduzco, C.; Sánchez-Escudero, J. Purslane (*Portulaca oleracea* L.) response to NPK fertilization. *Rev. Fitotec. Mex.* **2017**, *40*, 325–332. [CrossRef]

44. Petropoulos, S.; Karkanis, A.; Fernandes, Â.; Barros, L.; Ferreira, I.C.F.R.; Ntatsi, G.; Petrotos, K.; Lykas, C.; Khah, E. Chemical composition and yield of six genotypes of common purslane (*Portulaca oleracea* L.): An alternative source of omega-3 fatty acids. *Plant Foods Hum. Nutr.* **2015**, *70*, 420–426. [CrossRef]
45. Karkanis, A.C.; Fernandes, A.; Vaz, J.; Petropoulos, S.; Georgiou, E.; Ciric, A.; Sokovic, M.; Oludemi, T.; Barros, L.; Ferreira, I. Chemical composition and bioactive properties of *Sanguisorba minor* Scop. under Mediterranean growing conditions. *Food Funct.* **2019**, *10*, 1340–1351. [CrossRef]
46. Finimundy, T.C.; Karkanis, A.; Fernandes, Â.; Petropoulos, S.A.; Calhelha, R.; Petrović, J.; Soković, M.; Rosa, E.; Barros, L.; Ferreira, I.C.F.R. Bioactive properties of *Sanguisorba minor* L. cultivated in central Greece under different fertilization regimes. *Food Chem.* **2020**, *327*, 127043. [CrossRef]
47. Montesano, F.F.; Gattullo, C.E.; Parente, A.; Terzano, R.; Renna, M. Cultivation of potted sea fennel, an emerging mediterranean halophyte, using a renewable seaweed-based material as a peat substitute. *Agriculture* **2018**, *8*, 96. [CrossRef]
48. Sarrou, E.; Siomos, A.S.; Riccadona, S.; Aktsoglou, D.C.; Tsouvaltzis, P.; Angeli, A.; Franceschi, P.; Chatzopoulou, P.; Vrhovsek, U.; Martens, S. Improvement of sea fennel (*Crithmum maritimum* L.) nutritional value through iodine biofortification in a hydroponic floating system. *Food Chem.* **2019**, *296*, 150–159. [CrossRef]
49. Renna, M.; Gonnella, M.; Caretto, S.; Mita, G.; Serio, F. Sea fennel (*Crithmum maritimum* L.): From underutilized crop to new dried product for food use. *Genet. Resour. Crop Evol.* **2017**, *64*, 205–216. [CrossRef]
50. Bonasia, A.; Lazzizera, C.; Elia, A.; Conversa, G. Nutritional, biophysical and physiological characteristics of wild rocket genotypes as affected by soilless cultivation system, salinity level of nutrient solution and growing period. *Front. Plant Sci.* **2017**, *8*, 300. [CrossRef]
51. Petropoulos, S.A.; Fernandes, Â.; Dias, M.I.; Pereira, C.; Calhelha, R.C.; Ivanov, M.; Sokovic, M.D.; Ferreira, I.C.F.R.; Barros, L. Effects of growing substrate and nitrogen fertilization on the chemical composition and bioactive properties of *Centaurea raphanina* ssp. *mixta* (DC.) Runemark. *Agronomy* **2021**, *11*, 576. [CrossRef]
52. Polyzos, N.; Paschoalinotto, B.; Compocholi, M.; Dias, M.I.; Barros, L.; Petropoulos, S.A. The effects of fertilization regime on growth parameters and bioactive properties of pot grown *Cichorium spinosum* L. *Biol. Life Sci. Forum* **2022**, *16*, 6.
53. Freitas, J.A.; Ccana-Ccapatinta, G.V.; Da Costa, F.B. LC-MS metabolic profiling comparison of domesticated crops and wild edible species from the family Asteraceae growing in a region of São Paulo state, Brazil. *Phytochem. Lett.* **2021**, *42*, 45–51. [CrossRef]
54. Rahimi, A.; Kamali, M. Different planting date and fertilizing system effects on the seed yield, essential oil and nutrition uptake of Milk thistle (*Silybum marianum* (L.) Gaertn). *Adv. Environ. Biol.* **2012**, *6*, 1789–1796.
55. Fontana, E.; Hoeberechts, J.; Nicola, S.; Cros, V.; Palmegiano, G.B.; Peiretti, P.G. Nitrogen concentration and nitrate ammonium ratio affect yield and change the oxalic acid concentration and fatty acid profile of purslane (*Portulaca oleracea* L.) grown in a soilless culture system. *J. Sci. Food Agric.* **2006**, *86*, 2417–2424. [CrossRef]
56. García-Herrera, P.; Sánchez-Mata, M.C.; Cámara, M.; Fernández-Ruiz, V.; Díez-Marqués, C.; Molina, M.; Tardío, J. Nutrient composition of six wild edible Mediterranean Asteraceae plants of dietary interest. *J. Food Compos. Anal.* **2014**, *34*, 163–170. [CrossRef]
57. Karik, U. The effect of different harvest dates on the yield and quality of the golden thistle (*Scolymus hispanicus* L.). *Turk. J. Field Crop.* **2019**, *24*, 230–236. [CrossRef]
58. Pieroni, A.; Janiak, V.; Dürr, C.M.; Lüdeke, S.; Trachsel, E.; Heinrich, M. In vitro antioxidant activity of non-cultivated vegetables of ethnic Albanians in southern Italy. *Phytother. Res.* **2002**, *16*, 467–473. [CrossRef]
59. Papadimitriou, D.M.; Daliakopoulos, I.N.; Kontaxakis, E.; Sabathianakis, M.; Manios, T.; Savvas, D. Effect of moderate salinity on Golden Thistle (*Scolymus hispanicus* L.) grown in a soilless cropping system. *Sci. Hortic.* **2022**, *303*, 111182. [CrossRef]
60. Papadimitriou, D.; Kontaxakis, E.; Daliakopoulos, I.; Manios, T.; Savvas, D. Effect of N:K ratio and electrical conductivity of nutrient solution on growth and yield of hydroponically grown golden thistle (*Scolymus hispanicus* L.). *Proceedings* **2020**, *30*, 87. [CrossRef]
61. Petropoulos, S.; Fernandes, Â.; Karkanis, A.; Antoniadis, V.; Barros, L.; Ferreira, I. Nutrient solution composition and growing season affect yield and chemical composition of *Cichorium spinosum* plants. *Sci. Hortic.* **2018**, *231*, 97–107. [CrossRef]
62. AOAC. *Official Methods of Analysis of Association of Official Analytical Chemists*; Horwitz, W., Latimer, G., Eds.; AOAC Inter.: Gaithersburg, MD, USA, 2016; ISBN 0935584773.
63. Barros, L.; Pereira, C.; Ferreira, I.C.F.R. Optimized analysis of organic acids in edible mushrooms from Portugal by Ultra-Fast Liquid Chromatography and Photodiode Array Detection. *Food Anal. Methods* **2013**, *6*, 309–316. [CrossRef]
64. Klados, E.; Tzortzakis, N. Effects of substrate and salinity in hydroponically grown *Cichorium spinosum*. *J. Soil Sci. Plant Nutr.* **2014**, *14*, 211–222. [CrossRef]
65. Di Mola, I.; Cozzolino, E.; Ottaiano, L.; Nocerino, S.; Rouphael, Y.; Colla, G.; El-Nakhel, C.; Mori, M. Nitrogen use and uptake efficiency and crop performance of baby spinach (*Spinacia oleracea* L.) and Lamb's Lettuce (*Valerianella locusta* L.) grown under variable sub-optimal N regimes combined with plant-based biostimulant application. *Agronomy* **2020**, *10*, 278. [CrossRef]
66. Karkanis, A.C.; Petropoulos, S.A. Physiological and growth responses of several genotypes of common purslane (*Portulaca oleracea* L.) under Mediterranean semi-arid conditions. *Not. Bot. Horti Agrobot. Cluj-Napoca* **2017**, *45*, 569–575. [CrossRef]
67. Petropoulos, S.A.; Fernandes, Â.; Dias, M.I.; Pereira, C.; Calhelha, R.C.; Ivanov, M.; Sokovic, M.D.; Ferreira, I.C.F.R.; Barros, L. The effect of nitrogen fertigation and harvesting time on plant growth and chemical composition of *Centaurea raphanina* subsp. *mixta* (DC.) Runemark. *Molecules* **2020**, *25*, 3175. [CrossRef]
68. Zhang, Y.P.S.Y.; Lin, X.Y.; Zhang, Y.P.S.Y.; Zheng, S.J.; Du, S.T. Effects of nitrogen levels and nitrate/ammonium ratios on oxalate concentrations of different forms in edible parts of spinach. *J. Plant Nutr.* **2005**, *28*, 2011–2025. [CrossRef]

69. Dias, M.I.; Barros, L.; Dueñas, M.; Pereira, E.; Carvalho, A.M.; Alves, R.C.; Oliveira, M.B.P.P.; Santos-Buelga, C.; Ferreira, I.C.F.R. Chemical composition of wild and commercial *Achillea millefolium* L. and bioactivity of the methanolic extract, infusion and decoction. *Food Chem.* **2013**, *141*, 4152–4160. [CrossRef]
70. Le Roux, M.R.; Khan, S.; Valentine, A.J. Organic acid accumulation may inhibit N2 fixation in phosphorus-stressed lupin nodules. *New Phytol.* **2008**, *177*, 956–964. [CrossRef]
71. Rietra, R.P.J.J.; Heinen, M.; Dimkpa, C.O.; Bindraban, P.S. Effects of nutrient antagonism and synergism on yield and fertilizer use efficiency. *Commun. Soil Sci. Plant Anal.* **2017**, *48*, 1895–1920. [CrossRef]
72. Aboyeji, C.M.; Dunsin, O.; Adekiya, A.O.; Suleiman, K.O.; Chinedum, C.; Okunlola, F.O.; Joseph, A.; Ejue, S.W.; Adesola, O.O.; Olofintoye, T.A.J.; et al. Synergistic and antagonistic effects of soil applied P and Zn fertilizers on the performance, minerals and heavy metal composition of groundnut. *Open Agric.* **2020**, *5*, 1–9. [CrossRef]
73. Panchal, P.; Miller, A.J.; Giri, J. Organic acids: Versatile stress-response roles in plants. *J. Exp. Bot.* **2021**, *72*, 4038–4052. [CrossRef]
74. Malvi, U.R. Interaction of micronutrients with major nutrients with special reference to potassium. *Karnataka J. Agric. Sci.* **2011**, *24*, 106–109.
75. Marmouzi, I.; El Karbane, M.; El Hamdani, M.; Kharbach, M.; Naceiri Mrabti, H.; Alami, R.; Dahraoui, S.; El Jemli, M.; Ouzzif, Z.; Cherrah, Y.; et al. Phytochemical and pharmacological variability in golden thistle functional parts: Comparative study of roots, stems, leaves and flowers. *Nat. Prod. Res.* **2017**, *31*, 2669–2674. [CrossRef] [PubMed]
76. Petropoulos, S.; Ntatsi, G.; Levizou, E.; Barros, L.; Ferreira, I. Nutritional profile and chemical composition of *Cichorium spinosum* ecotypes. *LWT Food Sci. Technol.* **2016**, *73*, 95–101. [CrossRef]
77. Hussain, N.; Li, H.; Jiang, Y.; Jabeen, Z.; Shamsi, I.H.; Ali, E.; Jiang, L. Response of seed tocopherols in oilseed rape to nitrogen fertilizer sources and application rates. *J. Zhejiang Univ. Sci. B* **2014**, *15*, 181–193. [CrossRef] [PubMed]
78. Cocetta, G.; Casciani, D.; Bulgari, R.; Musante, F.; Kolton, A.; Rossi, M.; Ferrante, A. Light use efficiency for vegetables production in protected and indoor environments. *Eur. Phys. J. Plus* **2017**, *132*, 43. [CrossRef]
79. Rosales, M.A.; Cervilla, L.M.; Sánchez-Rodríguez, E.; Rubio-wilhelmi, M.; Blasco, B.; Ríos, J.J.; Soriano, T.; Castilla, N.; Romero, L.; Ruiz, J.M. The effect of environmental conditions on nutritional quality of cherry tomato fruits: Evaluation of two experimental Mediterranean greenhouses. *J. Sci. Food Agric.* **2011**, *91*, 152–162. [CrossRef]
80. Fageria, N.K.; Oliveira, J.P. Nitrogen, phosphorus and potassium interactions in upland rice. *J. Plant Nutr.* **2014**, *37*, 1586–1600. [CrossRef]
81. Tränkner, M.; Tavakol, E.; Jákli, B. Functioning of potassium and magnesium in photosynthesis, photosynthate translocation and photoprotection. *Physiol. Plant.* **2018**, *163*, 414–431. [CrossRef]
82. Xie, K.; Cakmak, I.; Wang, S.; Zhang, F.; Guo, S. Synergistic and antagonistic interactions between potassium and magnesium in higher plants. *Crop J.* **2021**, *9*, 249–256. [CrossRef]
83. Morales, P.; Ferreira, I.C.F.R.; Carvalho, A.M.; Sánchez-Mata, M.C.; Cámara, M.; Tardío, J. Fatty acids profiles of some Spanish wild vegetables. *Food Sci. Technol. Int.* **2012**, *18*, 281–290. [CrossRef]

horticulturae

MDPI

Article

Changes in Carbohydrates, Organic Acids, and Minerals at Different Development Stages of *Hexachlamys edulis* Fruit, a Wild South American Species with Horticultural Potential

Miriam Elisabet Arena [1,*], Ignacio Sebastián Povilonis [1], Virginia Borroni [2], Ethel Pérez [3], Néstor Pellegrino [4], Claudio Cacciatore [5] and Silvia Radice [1]

1. Laboratorio de Fisiología Vegetal, Machado 914, CONICET—Universidad de Morón, Morón B1708EOH, Argentina
2. Facultad de Arquitectura, Diseño y Urbanismo, Instituto de Tecnología en Polímeros y Nanotecnología, Universidad de Buenos Aires—CONICET, Ciudad Autónoma de Buenos Aires 1428, Argentina
3. Planta Piloto de Ingeniería Química (PLAPIQUI), Universidad Nacional del Sur, Consejo Nacional de Investigaciones Científicas y Técnicas (UNS-CONICET), Bahía Blanca 8000, Argentina
4. Facultad de Farmacia y Bioquímica, Universidad de Buenos Aires, Ciudad Autónoma de Buenos Aires 1113, Argentina
5. Escuela Superior de Ingeniería, Informática y Ciencias Agroalimentarias, Universidad de Morón, Morón B1708EOH, Argentina

* Correspondence: miriamearena@gmail.com

Abstract: The aim of this work was to study the patterns of the accumulation of carbohydrates, organic acids, and minerals at different development stages of *Hexachlamys edulis* fruit for its evaluation as a source of health-promoting compounds, which is necessary in order to be included in the Argentine Food Code. Additionally, the obtained results will allow for deciding the optimal time for consumption to receive a better flavour and a good contribution of the nutrients evaluated. The succinic acid concentration (the major organic acid) was high in unripe fruit (112.33 mg/g of the dry weight), then decreased to a minimum in medium ripe and ripe fruit (92.48 to 99.43 mg/g of the dry weight,), to increase again in overripe fruit (115.65 mg/g of the dry weight,). Sucrose increased significantly from 21.20 mg/g of the dry weight in unripe fruit to a maximum of 82.53 mg/g of the dry weight in ripe fruit. Glucose increased significantly from 95.59 mg/g of the dry weight in unripe fruit to a maximum of 163.13 mg/g of the dry weight in overripe fruit. Fructose followed the same behaviour, increasing significantly from 150.08 mg/g of the dry weight in unripe fruit to a maximum of 205.85 mg/g of the dry weight in overripe fruit. The starch concentration was at the maximum in unripe and medium ripe fruit (171.39 and 161.19 mg starch/g of the dry weight, respectively), to then decrease in ripe and overripe fruit (40.45 and 65.96 mg starch/mg of the dry weight, respectively). Maximum insoluble dietary fibre values were attained in unripe and medium ripe fruit (26.71 and 27.13 mg/100 g of the dry weight, respectively), to then decrease in ripe and overripe fruit (15.81 and 15.51 mg/100 g of the dry weight, respectively). Soluble dietary fibre oscillated between 9.03 and 11.26 mg/100 g of the dry weight during the development stages, although without significant differences. The mineral concentrations (Mg, K, Mn, and total cations) did not vary significantly during the different development stages. The obtained results allow us to consider *H. edulis* fruit as a promising natural source of sugars, organic acids, and minerals.

Keywords: underutilised fruit; fructose; glucose; dietary fibre; succinic acid

Citation: Arena, M.E.; Povilonis, I.S.; Borroni, V.; Pérez, E.; Pellegrino, N.; Cacciatore, C.; Radice, S. Changes in Carbohydrates, Organic Acids, and Minerals at Different Development Stages of *Hexachlamys edulis* Fruit, a Wild South American Species with Horticultural Potential. *Horticulturae* **2023**, *9*, 314. https://doi.org/10.3390/horticulturae9030314

Academic Editors: Roberta Bulgari, Ada Baldi, Anna Lenzi and Antonios Chrysargyris

Received: 30 January 2023
Revised: 17 February 2023
Accepted: 22 February 2023
Published: 28 February 2023

1. Introduction

Fruits with market value but which are rarely available to consumers due to a lack of cultivation are known as underutilised fruits. Hence, most of them are yet wild or semi-domesticated. Likewise, they are related to the local culture and have been part of ancestral food and medicine. In general, they overlap or are located in areas close to traditional crops and are neglected by agricultural research organisations [1]. Many of them are resistant to biotic and abiotic factors and retain desirable genes which could be useful in crop improvement through genetic engineering [2]. Given their composition in bioactive compounds, fatty acid profile, organic acids, and carbohydrates, underutilised fruits are not only considered as a food source but also for their therapeutic potential [3]. So, in some underutilised fruits such as *Euterpe oleracea*, Andean berries, and *Myrciaria dubia*, the positive effect of the bioactive substances on the antioxidant status and oxidative stress in humans was verified [4]. Nowadays, consumers are aware of the nutritional value of the new products they incorporate into their diet. Many of the tropical and temperate climate indigenous fruits are still underexploited due to a lack of knowledge, market conditions, and crop yield. While some 3,000 species of tropical fruits make up the diversity, only a few are cultivated on a large scale [5,6].

Hexachlamys edulis (O. Berg) Kausel and D. Legrand, "ubajay", is an underutilised species native from South America belonging to the Myrtaceae family. From the Atlantic Forest, it extends along the Paraná and Uruguay rivers. There are even references in the Paraná Jungle and the Paraguay river. In Argentina, it is found in the provinces of Entre Ríos, Corrientes, Misiones, Santa Fe, Formosa, and Chaco. *H. edulis* is a fruit tree with yellow globose drupes which are sweet–sour to very acidic, pleasant, and which quickly overripen. Fruit set and ripening occur during spring to early summer in South America. This species is undoubtedly prominent due to all its potential uses based on its nutraceutical properties, particularly in the leaves and in the yellow fruits rich in carotenoids [7,8]. The positive effects on the health of its leaves are well known. Bronchitis, cough, and whooping cough are all treated in alternative medicine with infusions of ubajay leaves, such as tea or mate. Many pharmacologic properties of leaf extracts are due to its powerful antioxidant action due to its flavonoids and tannins-rich content. Their effects on the balance of blood glucose levels for the treatment of diabetes and the reduction in hyperuricemia have been observed [7]. The consumption of carotenoid compounds has positive health impacts, including defence against cancer, cardiovascular disease, age-related disorders, and oxidative stress, as well as protection against macular degeneration [9]. The ripe fruit stands out for its antioxidant activity (near 100% DPPH radical scavenging activity with 5 mg methanolic fruit extract/mL) and nutraceutical profile, with a high content of carotenoids (706 µg of total carotenoids/g of the dry weight), making lutein the most abundant carotenoid [8]. Additionally, several efforts have been made to prospect and study the variability of this species in Entre Ríos for its subsequent selection and improvement [10] in its behaviour in other exotic environments and its reproductive phenological development regime [11], as well as in its post-harvest behaviour for its conservation [12].

Organoleptic characteristics are important determinants of consumer acceptability and, therefore, to the market possibilities of a fruit. In addition to colour and texture, flavour changes during fruit ripening due to variations in the ratio between organic acids (sourness) and sugar (sweetness) levels [8,13,14]. These two types of molecules are interconnected through the central carbon metabolism since both provides substrates for the respiratory processes. In addition, sugars and organic acids are intermediates in the biosynthesis of amino acids, vitamins, and terpenic aroma volatiles [13].

Various metabolite processes occur during fruit development and ripening, and products such as sugars and organic acids play an important role in developing fruit quality [15]. Internal and external factors determine the content of sugars, as well as their transport, metabolism, accumulation, and the relationship between them [16]. Glucose and fructose are the major simple sugars of fruits, and the relative amounts between them vary among fruits and in the same fruit in relation to maturity [17]. The health benefits of consuming dietary fibre can include things such as laxation, lowering blood sugar and cholesterol levels, and reducing the risk of developing colorectal cancer [18]. El-Zoghbi [19] has reported the variation in the dietary fibre content together with the fruit firmness during ripening.

Fruits have a variety of different organic acids in their fleshy parts, but the amount of each can vary a lot depending on the species and cultivars. Fruit flavour, to be consumed raw or used in fruit products, is influenced by the presence of organic acids in their fleshy parts. Malic, citric, isocitric, galacturonic, quinic, oxalic, and tartaric organic acids are very abundant in some fruits, while phenolic and ascorbic acids are ubiquitous in fruits [20].

The mineral nutrient fruit composition is important from a nutritional point of view, but it also can influence fruit ripening, quality, physiological disorders, storage behaviour, and other post-harvest physiological aspects (cell wall texture, non-enzymatic and enzymatic components of antioxidative system, plant/tissue defence, etc.) [21]. Different mineral elements such as sodium, potassium, iron, calcium, and many other trace elements are very essential for the human body [1]. There are several pre-harvest (genetics, plant density, temperature, soil composition, soil pH, and salt content) and post-harvest factors (harvesting time and gamma irradiation) that influence the contents of minerals in horticultural crops [21–24].

At this time, knowledge about the quality and potential use of wild fruits throughout the world is still scarce. Their biochemical analysis could provide information to breeders, producers, and consumers. The aim of this work was to study the patterns of the accumulation of carbohydrates, organic acids, and minerals during different stages of *H. edulis* fruit development for its evaluation as a source of health-promoting compounds, which is necessary in order to be included in the Argentine Food Code. Additionally, the obtained results will allow for deciding the optimal time for its consumption to achieve a better flavour and receive a good contribution of the nutrients evaluated.

2. Materials and Methods

2.1. Plant Material and Growing Conditions

Seventeen *Hexachlamys edulis* plants were obtained from the seeds of fruits collected in Federación, Entre Ríos (Argentina) (30°59′ SL, 57°55′ WL, 50 m.a.s.l.). They were grown in the nursery and then were transplanted to the experimental field of the University of Morón (Moreno, Buenos Aires, 34°35′4.98″ SL, 58°48′52.09″ WL, 14 m.a.s.l.). The mean air daily temperatures in October, November, and December 2018, months when the flowering and fruit growth and ripening took place, were 17.3, 20.9, and 22.0 °C, respectively. Rainfall was 200, 81, and 157 mm along the mentioned months, respectively. Fruits (3 samples of 15 fruits each one) were harvested during November and December 2018 in 4 development stages according to Arena et al. [8]. Stage 1: unripe fruits with all green skin (fruits with 21 days after full bloom: dafb); stage 2: medium ripe fruits with green and yellow skin (fruits with 35 dafb); stage 3: ripe fruits with all yellow skin (fruits with 42 dafb); and stage 4: overripe fruits with yellow and brown skin (fruits with 49 dafb) (Figure 1).

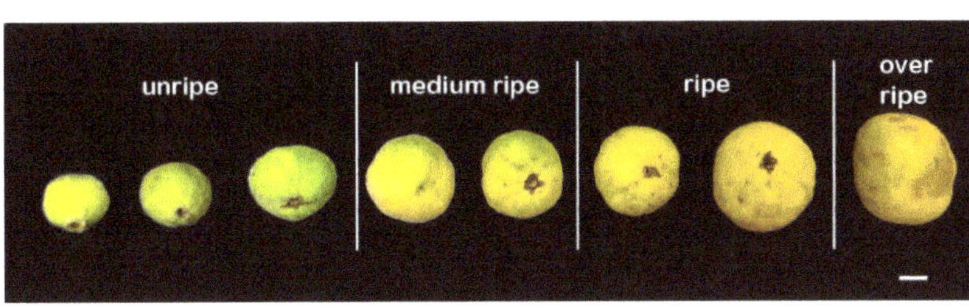

Figure 1. *H. edulis* fruit at different development stages. Bar = 1 cm.

2.2. Analysis of Carbohydrates and Organic Acids

Immediately after harvest, the fruits were lyophilised and stored at −20 °C until analysis. The carbohydrate and organic acid extraction was achieved according to Colaric et al. [25]. An aliquot of the extract was filtered through a 0.45 μm MILLIPORE filter. A Waters e2695 HPLC system (Waters Associates, Milford, MA, USA) equipped with a refraction index detector (Waters 2414) was used. Carbohydrates were separated on Rezex RCM—Monosaccharide Ca^{2+} (8%) (300 mm length, 7.8 mm i.d., Phenomenex, Torrance, CA, USA). The operating conditions recommended by the supplier were used: the mobile phase included ultrapure water; a flow rate of 0.6 mL/min; and a column temperature of 65 ± 5 °C, and the injection volume loop was 10 μL. They were quantified by the external standard method using the reference standards of glucose, fructose, and sucrose (Anedra, Buenos Aires, Argentina). Then, the sum of the three quantified sugars was calculated (total sugars). Carbohydrate contents were expressed as mg per g of the dried weight. Additionally, the glucose/fructose ratio was calculated.

The separation of organic acids were carried out by Rezex ROA—Organic Acid H+ (8%) column (300 × 7.8 mm) (Bio-Rad, Hercules, CA, USA), associated with a photodiode array detector at 210 nm (Waters 2998). The HPLC equipment was the same as mentioned above. The column operated at 65 ± 5 °C. The sulfuric solution (0.005N) was a mobile phase at a flow rate of 0.6 mL/min. Tartaric, malic, quinic, and succinic acids were quantified by an external standard method. Then, the sum of the four quantified organic acids was calculated (total organic acids). Concentrations of the quantified organic acids were expressed as mg per g of the dry weight. In all cases, determinations were performed in duplicate. Additionally, the total sugars/total organic acids ratio was calculated.

Starch was quantified according to the methodology described by Lage-Yuste et al. [26]. Samples (0.2 g for unripe and medium ripe fruits, and 1.0 g for ripe and overripe fruits) were taken and homogenised with 5 mL of 52% perchloric acid and left to stand for 10 min. Then, the volume was completed to 100 mL, and 10 mL of this solution was taken out and mixed with 0.5 mL of 0.1:1 iodine:potassium iodide solution. The solutions were kept for 10 min in light–dark and then 1.5 mL of this solution were taken and centrifugated (10,000 rpm for 1 min). The absorbances were quantified at 600 nm in a Spectrum SP-2000 spectrophotometer. A calibration curve was prepared with starch (Sigma, Burlington, MA, USA) at 2, 3, 4, and 5 mg/100 mL, and the concentrations were reported as mg starch/g of the dry weight.

The determination of total, soluble, and insoluble dietary fibre was performed using the Dietary Fibre Assay Kit according to AOAC (991.43) and AACC (32.05.01) approved methods from Megazyme [27]. Fibre determination was performed in duplicate on samples of dry material. The gelatinisation, hydrolysis, and depolymerisation of starch were achieved by mixing the samples with heat-stable α-amylase at 100 °C. Then, the samples were incubated at 60 °C with protease (to solubilise and depolymerise proteins) and amyloglucosidase (to hydrolyse starch fragments to glucose) and they were treated with ethanol (rate 1:4, v/v) to precipitate the soluble fibre and remove the depolymerised protein and glucose (from starch). The residue was filtered, washed with 78% ethanol, 95% ethanol, and acetone, dried, and weighed. One duplicate was analysed for the protein content and the other was incubated at 525 °C to determine the ash. The total dietary fibre was determined by gravimetry based on the weight of the filtered residue less the weight of the protein and ash. For the soluble and insoluble fibre determination, before precipitation with 78% ethanol, the sample was filtered to obtain insoluble dietary fibre and then the residue was washed with warm distilled water. Filtered solution and rinse waters were collected and treated with 95% ethanol (rate 1:4 v/v) to precipitate the soluble dietary fibre. Precipitate was separated by being filtered and dried. Both the soluble dietary fibre and insoluble dietary fibre calculated values were corrected for the protein, ash, and blank content.

2.3. Analysis of Minerals

Approximately 1 g of each sample was calcined in a muffle at 500 °C for 5 h. The ash was raised with 5 mL of 20% HCl and completed to 50 mL with distilled water. Then, it was filtered and quantified by atomic absorption spectrometry in a Perkin Elmer AAnalyst 200 equipment. The minerals Mg, K, and Mn were quantified. Then, the sum of the three quantified minerals was calculated (total cations). Mg and K were selected because they are important minerals in the electrolytic balance, which present highlighted properties in populations of children, adults, and athletes, while was selected Mn for its antioxidant activity.

2.4. Statistical Analysis

Data were analysed through general linear models (GLM; mixed models) when corresponded. The assumption of normality was checked by the Shapiro–Wilk test and QQplots. Additionally, the homogeneity of variance was checked by the Levene test and scatter plots of the residuals versus the predicted values of each model. Data were analysed through an ANOVA and the means were separated by Tukey's test at $p \leq 0.05$ using RStudio. Pearson's correlation analysis was used to examine the relationship between the total titratable acidity [8] with each organic acid and with the total acids to understand how these variables were related.

3. Results and Discussion

The concentration of organic acids in the fleshy portions of the fruits can vary significantly depending on the species, environmental factors, the kind of the tissue, and the development stage of the fruit [20]. Usually, a decrease in the organic acid concentration and acidity is observed during fruit ripening. Examples of this behaviour are pears, peaches, and apples [28]. However, some fruit species show an increase in organic acids during ripening, as was demonstrated for "mangaba" fruit (*Hancornia speciosa*) and "soursop" (*Annona muricata*), which are considered bittersweet fruits [28,29].

Interestingly, the organic acid concentration varied significantly over the different development stages of *H. edulis* fruit (Table 1). The succinic acid concentration (the main organic acid) was high in unripe fruit (112.33 mg/g of the dry weight), then decreased to a minimum in medium ripe and ripe fruit (92.48 to 99.43 mg/g of the dry weight), to increase again in overripe fruit (115.65 mg/g of the dry weight). The malic acid concentration (the second main organic acid) was 44.90 mg/g of the dry weight in unripe fruit, to increase to a maximum of 105.14 mg/g of the dry weight in ripe fruit, after which it decreased again to 84.79 mg/g of the dry weight in overripe fruit. The quinic acid concentration was at the minimum in unripe fruit (0.37 mg/g of the dry weight), to then increase during fruit development (25.25 to 22.85 mg/g of the dry weight). Tartaric acid was detected only in overripe fruit (0.94 mg/g of the dry weight). Finally, the unripe fruits showed the minimum value for the total organic acids (157.67 mg/g of the dry weight) to increase during the fruit development between 207.98 and 227.27 mg/g of the dry weight (Figure 2).

Table 1. ANOVA for tartaric acid (TAR), malic acid (MAL), quinic acid (QUI), and succinic acid (SUC) expressed in dry weight and considering the four development stages of *H. edulis* "ubajay" harvested from the plants growing in Moreno (Buenos Aires). Values represent means ± standard error.

Factor	TAR (mg/g)	MAL (mg/g)	QUI (mg/g)	SUC (mg/g)
Stages				
Unripe	0.00 ± 0.10 b	44.90 ± 8.33 c	0.37 ± 0.24 b	112.33 ± 3.31 a
Medium Ripe	0.00 ± 0.10 b	90.19 ± 2.85 b	25.25 ± 1.80 a	92.48 ± 3.31 b
Ripe	0.00 ± 0.10 b	105.14 ± 1.95 a	22.62 ± 1.09 a	99.43 ± 3.31 b
Overripe	0.94 ± 0.10 a	84.79 ± 2.34 b	22.85 ± 1.78 a	115.65 ± 3.31 a
F	26.039	77.887	262.200	12.398
p	<0.001	<0.001	<0.001	<0.001

F(p) = F statistic and probability of Fisher test. Different letters in each column indicate significant differences according to the Tukey test ($p \leq 0.05$).

As in *H. edulis* fruits, succinic acid was the major organic acid present throughout most of the development stages of *Litchi chinensis* and *Amelanchier alnifolia* (saskatoon), although a higher malic acid concentration was found at maturity [30,31]. Similar to *H. edulis* fruits, in *Ziziphus jujuba*, the major components of organic acid were malic acid, quinic acid, and succinic acid, although malic acid was predominant [32]. In citrus fruits, quinic acid was the major organic acid at the beginning of the development; later, citric acid was predominant in acidic varieties, while in less acidic types, malic acid overtook it [33]. In apricot, plum, plumcot, and peach, quinic acid was the third predominant organic acid too, not showing a marked decrease in their concentrations [34], as was shown in *H. edulis* fruits.

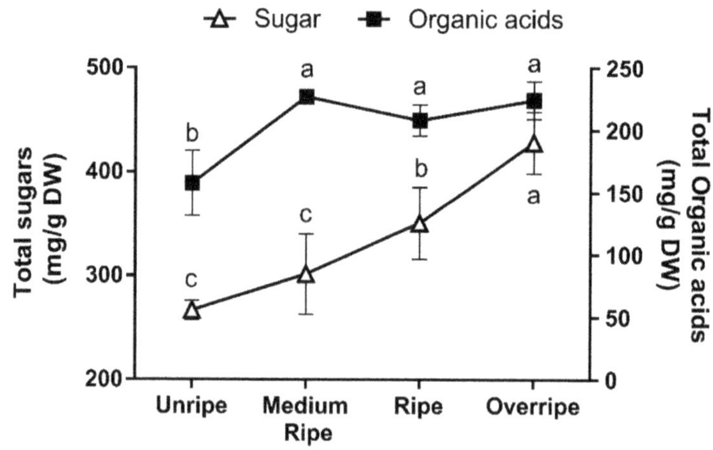

Figure 2. Total sugars and total organic acids (mg/g dry weight) (DW) during the development stages of *H. edulis* fruit: unripe, medium ripe, ripe, and overripe. Values represent means ± S.E. Different letters indicate significant differences according to the Tukey test ($p \leq 0.05$).

In *H. edulis* fruits, the total titratable acidity stayed high during unripe, medium ripe, and ripe stages, and decreased in the overripe stage (1.8, 1.6, 1.8, and 1.4%, respectively) [8] following different behaviour with respect to the total organic acids, a fact that explained the non-significant correlation found between both of these variables (r = −0.439, p = 0.1537). The only acid that decreased in the overripe stage compared with the ripe stage was malic acid, which showed a positive and significant correlation with the total acids (r = 0.930, $p < 0.001$), while quinic acid also presented a positive and significant correlation with the total acids (r = 0.855, $p < 0.001$). In both cases, these acids explained the increase in the total acid concentration throughout the fruit stages. In subtropical fruit *Annona cherimola*, a significant increase in the organic acid levels during ripening occurred as in *H. edulis*, and the increase in acidity was also related to the accumulation of malic acid [28,29]. The acid concentration in fruits affects the fruit's taste, as was cited for strawberries, blackberries, and mandarins [35].

The concentration of soluble sugars varies during fruit growth and development among species, usually peaking at maturity. Glucose and fructose predominate in the majority of fruits, whereas in others such as mandarin, peaches, and litchi, sucrose is the most important sugar [36]. The simple sugars varied significantly during the different development stages of *H. edulis* fruit (Table 2). Sucrose increased significantly from 21.20 mg/g of the dry weight in unripe fruit to a maximum of 82.53 mg/g of the dry weight in ripe fruit, to then it decreased to 59.38 mg/g of the dry weight in overripe fruit. Glucose increased significantly from 95.59 mg/g of the dry weight in unripe fruit to a maximum of 163.13 mg/g of the dry weight in overripe fruit. Fructose followed the same behaviour, increasing significantly from 150.08 mg/g of the dry weight in unripe fruit to a maximum of 205.85 mg/g of the dry weight in overripe fruit. The glucose concentration was lower than fructose. So, the glucose/fructose ratio increased significantly from 0.63 to 0.79 during fruit development. Finally, the total sugars in unripe fruits showed the minimum value (266.88 mg/g of the dry weight) to increase significantly during fruit development (301.68 to 428.36 mg/g of the dry weight) (Figure 2). The total sugars/total organic acids ratio was at the maximum in overripe fruit (1.90). The starch concentration was at the maximum in unripe and medium ripe fruit (171.39 and 161.19 mg starch/g of the dry weight, respectively), to then decrease in ripe and overripe fruit (40.45 and 65.96 mg starch/mg of the dry weight, respectively).

Table 2. ANOVA for sucrose (SUC), glucose (GLU), fructose (FRU), glucose/fructose ratio (GLU/FRU), total sugars/total organic acids (TS/TA), and starch (STR) expressed in dry weight and considering the four development stages of *H. edulis* "ubajay" harvested in Moreno (Buenos Aires). Values represent means ± standard error.

Factor	SUC (mg/g)	GLU (mg/g)	FRU (mg/g)	GLU/ FRU	TS/TA	STR (mg/g)
Stages						
Unripe	21.20 ± 4.80 b	95.59 ± 2.42 b	150.08 ± 6.50 bc	0.63 ± 0.01 b	1.72 ± 0.09 ab	171.39 ± 1.93 a
Medium Ripe	77.79 ± 4.80 a	89.49 ± 5.90 b	134.40 ± 6.50 c	0.66 ± 0.00 b	1.44 ± 0.09 b	161.19 ± 1.93 a
Ripe	82.53 ± 4.80 a	105.00 ± 77.59 b	163.08 ± 6.50 b	0.64 ± 0.00 b	1.54 ± 0.09 ab	40.45 ± 1.93 b
Overripe	59.38 ± 4.80 a	163.13 ± 6.57 a	205.85 ± 6.50 a	0.79 ± 0.00 a	1.90 ± 0.09 a	65.96 ± 1.93 b
F	33.672	42.000	33.660	118.100	4.972	11.705
p	<0.001	<0.001	<0.001	<0.001	0.045	0.003

F(p) = F statistic and probability of Fisher test. Different letters in each column indicate significant differences according to the Tukey test ($p \leq 0.05$).

As in the overripe fruit of *H. edulis*, the sucrose decrease was concomitant to an increase in the glucose and fructose levels during peach (*Prunus persica*) ripening, suggesting the degradation of this disaccharide [36]. In addition, starch degradation could contribute to the increase in the glucose content, as it was demonstrated for *Malus domestica Borkh.* cv. Gala (apples) and *Actinidia deliciosa* (kiwifruit) [37,38].

In *H. edulis* fruit, the main sugars present are fructose and glucose, as occurs in grapes, making the fructose concentration higher than the glucose concentration. In tomato, melon, grape berry, cherry, and peach, fructose and glucose are found in identical quantities. However, in apple, fructose is the major sugar [39]. The sweetness of fructose is higher than that of glucose, so changes in these sugar concentrations affect the sweet taste of the fruits as it occurs in grapes [40]. In addition, sugars are the main precursor of aroma in fruits [41]. The amount of carbohydrates in *H. edulis* ripe fruit (39.2 mg/100 g of the fresh weight) were higher than those reported for several underutilised species from India (0.50 to 34.40 mg/100 g of the fresh weight for *Carissa carandas* and *Terminalia belirica*, respectively) [6].

The decrease in the starch content in *H. edulis* fruit with the development stages has also been observed in several climacteric fruits, such as in tomato (10–20 dry weight to 0.1 mg/g of the dry weight), banana (100–300 to <150 mg/g of the dry weight), apple (20–25 to 0.5 mg/g of the dry weight), and mango (60 to 5 mg/g of the dry weight). This behaviour was not observed in non-climacteric fruits, where the starch content decreases sharply after anthesis and therefore the fruits accumulate simple sugars during development. Although many reports show the differences in the starch metabolism between climacteric and non-climacteric fruits, it is unclear if this contrast could be a key to differentiate both classes of fruits [42,43].

Interestingly, Colaric et al. [25] found that the sweetness of nectarine and peach was positively correlated with a higher sugars/organic acid ratio more than to the total amount of sugars alone. Additionally, in strawberries, a high sugar/organic acid ratio was associated with a strong sweetness and weak sourness [44]. In addition, acidity was related to the concentrations and type of the different organic acids, which are ordered according to its sourness relative to citric acid, as follows: citric (1.0) > malic (0.9) > tartaric (0.8) [45]. It is reported that malic acid in particular provides a smooth, tart taste. Taste was related to the malic/citric acid ratio, total sugars, sucrose, sorbitol, and malic acid concentrations in nectarine and peaches [25,46]. Therefore, the increase in the total sugar, together with the lack of changes in the total organic acids at the ripe and overripe stages, suggest a more sweet and less sour fruit. These results suggest that in order to obtain an appealing product, ubajay fruits should be consumed in a ripe or overripe state, although the overripe fruits presented lower antioxidant activity with respect to ripe fruits [8]. Additionally, considering that the post-harvest life of the overripe fruit could be short, the ripe stage is desirable to extend the time for consumption.

Maximum insoluble dietary fibre values were attained in unripe and medium ripe fruit (26.71 and 27.13 mg/100 g of the dry weight, each), to then decrease in ripe and overripe fruits to 15.81 and 15.51 mg/100 g of the dry weight, respectively (Table 3). Soluble dietary fibre oscillated between 9.03 and 11.26 mg/100 g of the dry weight during the development stages, although without significant differences. So, the total dietary fibre decreased from 36.80 to 25.94 mg/100 g of the dry weight between medium ripe and overripe fruit. The pattern of change in the amount of dietary fibre with ripening found in *H. edulis* is similar to other tropical fruits, where the decrease in the total fibre content was associated with softening of the flesh [19]. Hydrolysis of the cell wall by indigenous cellulolytic and pectinolytic enzymes was responsible for this change [19]. The amounts of total fibre in *H. edulis* ripe fruit (3.03 mg/100 g of the fresh weight) were higher than those reported for *M. indica* and *P. guajava* (1.46 to 1.80 and 1.81 mg/100 g of the fresh weight, respectively) [19], and higher than those cited for several underutilised species from India (0.10 to 3.00 mg/100 g of the fresh weight for *Morus indica* and *Aegle marmelos*, respectively) [6]. Therefore, *H. edulis* may constitute a new source of fibre for the consumers.

Table 3. ANOVA for insoluble dietary fibre (IDF), soluble dietary fibre (SDF), and total dietary fibre (TDF) expressed in dry weight and considering the four development stages of *H. edulis* "ubajay" harvested in Moreno (Buenos Aires). Values represent means ± standard error.

Factor	IDF (g/100 g)	SDF (g/100 g)	TDF (g/100 g)
Stages			
Unripe	26.71 ± 1.68 a	9.03 ± 1.35	35.74 ± 2.22 ab
Medium Ripe	27.13 ± 1.93 a	9.67 ± 1.35	36.80 ± 2.22 a
Ripe	15.81 ± 1.93 b	11.26 ± 1.35	27.07 ± 2.22 ab
Overripe	15.51 ± 1.93 b	10.44 ± 1.35	25.94 ± 2.22 b
F	14.895	0.505	6.511
P	0.001	0.689	0.015

F(p) = F statistic and probability of Fisher test. Different letters in each column indicate significant differences according to the Tukey test ($p \leq 0.05$).

The mineral concentrations (Mg, K, Mn, and total cations) did not vary significantly across the development of *H. edulis* fruit (Table 4). The Mg and K concentrations in ripe fruit were 0.40 and 22.85 mg/g of the dry weight, respectively, while the total cations concentration was 23.27 mg/g of the dry weight. Mn was absent in unripe, ripe and overripe fruits. In some underutilised fruits from India, Barua et al. [6] mentioned the absence of K in *Carissa carandas* and *Morus alba* fruits, while higher contents in *Terminalia chebula* fruit pulp (1270 mg/100 g of pulp) were cited. In different varieties of dates (*Phoenix dactylifera L.*), the K content was between 533.9 ± 0.95 and 1013 ± 0.86 mg/100 g and the Mg content was between 30.46 ± 0.40 and 76.74 ± 0.52 mg/100 g [47]. Czech et al. [48] studied different citrus fruits and reported that the fresh pulp contained 104–145 mg/100 g of K and 7.99–19.40 mg/100 g of Mg. Therefore, the mineral content of *H. edulis* is comparable with other fruits. K acquired through diet reduces the arterial pressure and the risk of stroke and coronary heart disease in adults [49]. On the other hand, Mg is a crucial mineral that works as a cofactor of different enzymes involved in antioxidant defences, the glucose metabolism, and blood pressure regulation. Therefore, increasing the K and Mg intake improves the cardiovascular function in adults [50]. Due to the importance for human health, the WHO recommends a minimum K level of 3510 mg/day and a Mg intake of 400 mg/day from food, depending on age and sex [51,52]. Considering that each fruit of ripe *H. edulis* is expected to contain ~140 mg of K and ~3.4 mg of Mg, the consumption of a portion of 5 fruits would incorporate ~16.0% and ~2.5% of the recommended intake of K and Mg, respectively.

Table 4. ANOVA for magnesium (Mg), potassium (K), manganese (Mn), and total cations (TC) expressed in dry weight and considering the four development stages of *H. edulis* "ubajay" harvested from the plants growing in Moreno (Buenos Aires). Values represent means ± standard error.

Factor	Mg (mg/g)	K (mg/g)	Mn (mg/g)	TC (mg/g)
Stages				
Unripe	0.33 ± 0.10	20.63 ± 7.63	0.00 ± 0.00	20.97 ± 7.74
Medium Ripe	0.34 ± 0.03	20.69 ± 2.55	0.01 ± 0.01	21.05 ± 2.57
Ripe	0.40 ± 0.02	22.85 ± 2.73	0.00 ± 0.00	23.27 ± 2.76
Overripe	0.24 ± 0.04	17.57 ± 3.14	0.00 ± 0.00	17.82 ± 3.17
F	3.559	0.761	2.815	0.787
p	0.087	0.556	0.130	0.543

F(p) = F statistic and probability of Fisher test.

4. Conclusions

Variations in carbohydrates, organic acids, and minerals through the development stages for *Hexachlamys edulis* fruit were analysed for the first time. The increase in the total sugar, together with the lack of changes in the total organic acids with the ripening process, suggest a more sweet and less sour fruit, i.e., an appealing product to be consumed at the ripe or overripe stage. Additionally, *H. edulis* may constitute a new source of fibre for consumers; additionally, the consumption of a portion of 5 ripe fruits would incorporate ~16% and ~2.5% of the recommended intake of K and Mg, respectively. These results enable us to propose *H. edulis* fruit as a promising natural source of sugars, organic acids, and minerals, information that is relevant for the introduction of *H. edulis* fruits into the Argentine Food Code.

Author Contributions: Conceptualization, M.E.A., I.S.P., V.B., E.P. and S.R.; Methodology, M.E.A., I.S.P., V.B., E.P., N.P., C.C. and S.R.; Software, I.S.P.; Validation, M.E.A. and I.S.P.; Formal analysis, M.E.A., I.S.P., E.P., N.P. and C.C.; Investigation, M.E.A., I.S.P., V.B., E.P., N.P., C.C. and S.R.; Resources, M.E.A., I.S.P., V.B., E.P., N.P., C.C. and S.R.; Data curation, M.E.A. and I.S.P.; Writing—original draft, M.E.A., I.S.P., V.B., E.P., N.P., C.C. and S.R.; Writing—review & editing, M.E.A., I.S.P., V.B., E.P., N.P., C.C. and S.R.; Visualization, M.E.A., I.S.P., V.B., E.P. and S.R.; Supervision, M.E.A. and S.R.; Project administration, M.E.A.; Funding acquisition, M.E.A. and S.R. All authors have read and agreed to the published version of the manuscript.

Funding: This research was partially funded by MINCYT-UM PICTO 2019-00003, and CONICET PIP 11220200102292CO. The APC did not receive external funding.

Data Availability Statement: Data is contained within the article.

Acknowledgments: The authors thank Marta Alonso for his technical assistance.

Conflicts of Interest: The authors declare no conflict of interest.

References

1. Hossain, M.M.; Rahim, M.A.; Haque, M.R. Biochemical properties of some important underutilized minor fruits. *J. Agric. Food Res.* **2021**, *5*, 100148. [CrossRef]
2. Murthy, H.N.; Bapat, V.A. Importance of Underutilized Fruits and Nuts. In *Bioactive Compounds in Underutilized Fruits and Nuts*; Murthy, H., Bapat, V., Eds.; Reference Series in Phytochemistry; Springer: Cham, Switzerland, 2020. [CrossRef]
3. Lamani, S.; Anu-Appaiah, K.A.; Murthy, H.N.; Dewir, Y.H.; Rikisahedew, J.J. Analysis of Free Sugars, Organic Acids, and Fatty Acids of Wood Apple (*Limonia acidissima* L.) Fruit Pulp. *Horticulturae* **2022**, *8*, 67. [CrossRef]
4. Gordon, A. Bioactive Compounds in Underutilized Tropical Fruits from Latin America. Ph.D. Dissertation, Universitäts-und Landesbibliothek, Bonn, Germany, 2012. Available online: https://nbn-resolving.org/urn:nbn:de:hbz:5n-29864 (accessed on 2 December 2022).
5. Nandal, U.; Bhardwaj, R.L. The role of underutilized fruits in nutritional and economic security of tribals: A review. *Crit. Rev. Food Sci. Nutr.* **2014**, *54*, 880–890. [CrossRef] [PubMed]
6. Barua, U.; Das, R.P.; Gogoi, B.; Baruah, S.R. Underutilized fruits of Assam for livelihood and nutritional security. *Agric. Rev.* **2019**, *40*, 175–184. [CrossRef]

7. Povilonis, I.S.; Arena, M.E.; Radice, S. *Hexachlamys edulis* (Berg) Kausel & Legrand, "ubajay", a native fruit species from South America. *AHS* **2021**, *35*, 389–397. [CrossRef]
8. Arena, M.E.; Povilonis, I.S.; Borroni, V.; Constenla, D.; Radice, S. Changes in physicochemical properties at different development stages of *Hexachlamys edulis* fruit, an underutilized South American species. *Heliyon* **2021**, *7*, e08323. [CrossRef]
9. Britton, G.; Khachik, F. Carotenoids in Food. In *Carotenoids: Nutrition and Health*; Britton, G., Pfander, H., Liaaen-Jensen, S., Eds.; Birkhäuser: Basel, Switzerland, 2009; Volume 5, pp. 45–66, ISBN 978-3-7643-7501-0. [CrossRef]
10. Povilonis, I.S.; Arena, M.E.; Radice, S. Caracterización de la variabilidad fenotípica de los frutos de ubajay (*Hexachlamys edulis*). In Proceedings of the 41° Congreso Argentino de Horticultura, La Plata, Argentina, 5–8 October 2021; Available online: http://sedici.unlp.edu.ar/bitstream/handle/10915/128494/Documento_completo.pdf-PDFA.pdf?sequence=1&isAllowed=y (accessed on 15 November 2022).
11. Radice, S.; Povilonis, I.S.; Arena, M.E. Flower and fruit formation of *Hexachlamys edulis* in Buenos Aires, Argentina. *JAEID* **2023**; *in press*.
12. Povilonis, I.S.; Arena, M.E.; Radice, S. Efecto de la temperatura de conservación sobre propiedades químicas de los frutos de *Hexachlamys edulis* en Concordia, Entre Ríos. In Proceedings of the VIII Congreso Internacional de Ciencia y Tecnología de Alimentos (CICYTAC), Cordoba, Argentina, 4–6 October 2022.
13. Batista-Silva, W.; Nascimento, V.L.; Medeiros, D.B.; Nunes-Nesi, A.; Ribeiro, D.M.; Zsögön, A.; Araújo, W.L. Modifications in Organic Acid Profiles during Fruit Development and Ripening: Correlation or Causation? *Front. Plant Sci.* **2018**, *9*, 1689. [CrossRef]
14. Shi, Y.; Li, B.J.; Su, G.; Zhang, M.; Grierson, D.; Chen, K.S. Transcriptional regulation of fleshy fruit texture. *J. Integr. Plant Biol.* **2022**, *64*, 1649–1672. [CrossRef]
15. Xie, F.; Chen, C.; Chen, J.; Yuan, Y.; Hua, Q.; Zhang, Z.; Zhao, J.; Hu, G.; Chen, J.; Qin, Y. Metabolic Profiling of Sugars and Organic Acids, and Expression Analyses of Metabolism-Associated Genes in Two Yellow-Peel Pitaya Species. *Plants* **2022**, *11*, 694. [CrossRef]
16. Zhi, W.S.; Li, Y.F.; Zi, S.W.; Xue, X.B. Sugar transport, metabolism, accumulation and their regulation in fruits. *J. Plant Physiol. Mol. Biol.* **2004**, *30*, 1–10. [PubMed]
17. Vicente, A.R.; Manganaris, G.A.; Sozzi, G.O.; Crisosto, C.H. Nutritional quality of fruits and vegetables. In *Postharvest Handling: A Systems Approach*, 2nd ed.; Florkowski, W.J., Shewfelt, R.L., Brueckner, B., Prussia, S.E., Eds.; 2009; pp. 57–106, ISBN 978-0-12-374112-7.
18. Ahmad, A.; Khalid, N. Chapter 5—Dietary Fibers in Modern Food Production: A Special Perspective with β-Glucans. In *Biopolymers for Food Design*; Grumezescu, A.M., Holban, A.M., Eds.; Academic Press: Philadelphia, PA, USA, 2018; pp. 125–156, ISBN 978-0-12-811449-0. [CrossRef]
19. El-Zoghbi, M. Biochemical changes in some tropical fruits during ripening. *Food Chem.* **1994**, *49*, 33–37. [CrossRef]
20. Walker, R.P.; Famiani, F. Organic Acids in Fruits: Metabolism, Functions and Contents. In *Horticultural Review*, 1st ed.; Warrington, I., Ed.; John Wiley & Sons, Ltd.: Hoboken, NJ, USA, 2018; pp. 371–430, ISBN 978-1-119-43107-7.
21. Paul, V.; Pandey, R.; Ramesh, K.V.; Singh, A. Role of mineral nutrients in physiology, ripening and storability of fruits. In *Advances in Plant Physiology*; Hemantaranjan, A., Ed.; An International Treatise Series: Nutriophysiological and Molecular Interventions for crop improvement under changing climate; Scientific Publishers: Jodhpur, India, 2012; Volume 13, pp. 56–96.
22. Rahman, M.M.; Roy, M.; Sajib, M.A.M.; Sarkar, A.; Hussain, M.S. Radiation effects on essential minerals content of cucumber (*Cucumis sativus*). *Amer. J. Food Nutr.* **2015**, *3*, 69–74.
23. Tyagi, S.; Sahay, S.; Imran, M.; Rashmi, K.; Mahesh, S.S. Pre-harvest factors influencing the postharvest quality of fruits: A review. *Curr. J. Appl. Sci. Technol.* **2017**, *23*, 12. [CrossRef]
24. Taghavi, T.; Siddiqui, R.; Rutto, L.K. The effect of preharvest factors on fruit and nutritional quality in strawberry. In *Strawberry Pre-and Post-harvest Management Techniques for Higher Fruit Quality*; IntechOpen: London, UK, 2019.
25. Colaric, M.; Stampar, F.; Solar, A.; Hudina, M. Influence of branch bending on sugar, organic acid and phenolic content in fruits of 'Williams' pears (*Pyrus communis* L.). *J. Sci. Food Agric.* **2006**, *86*, 2463–2467. [CrossRef]
26. Lage-Yusty, M.; Simal-Lozano, J.; Gómez, G.S. Determinación de almidón en alimentos. *Trab. Compostel. Biol.* **1983**, *10*, 15–25.
27. Megazyme. Total Dietary Fiber. Assay Procedure. 2017. Available online: https://www.megazyme.com/documents/Assay_Protocol/K-TDFR-200A_DATA.pdf (accessed on 10 June 2022).
28. González-Agüero, M.; Tejerina Pardo, L.; Zamudio, M.S.; Contreras, C.; Undurraga, P.; Defilippi, B.G. The Unusual Acid-Accumulating Behavior during Ripening of Cherimoya (*Annona cherimola* Mill.) is Linked to Changes in Transcription and Enzyme Activity Related to Citric and Malic Acid Metabolism. *Molecules* **2016**, *21*, 398. [CrossRef]
29. Carnelossi, M.A.G.; Costa de Sena, H.; Narain, N.; Yaguiu, P.; da Silva, G.F. Physico-Chemical Quality Changes in Mangaba (*Hancornia speciosa gomes*) Fruit Stored at Different Temperatures. *Braz. Arch. Biol. Technol.* **2009**, *52*, 985–990. [CrossRef]
30. Rogiers, S.Y.; Knowles, N.R. Physical and chemical changes during growth, maturation, and ripening of Saskatoon (*Amelanchier alnifolia*) fruit. *Can. J. Bot.* **1997**, *75*, 1215–1225. [CrossRef]
31. Pareek, S. Chapter 17—Nutritional and Biochemical Composition of Lychee (*Litchi chinensis* Sonn.) Cultivars. In *Nutritional Composition of Fruit Cultivars*; Simmonds, M.S.J., Preedy, V.R., Eds.; Academic Press: Cambridge, MA, USA, 2016; pp. 395–418, ISBN 9780124081178. [CrossRef]
32. Zhao, A.-L.; Xue, X.-F.; Wang, Y.-K.; Sui, C.-L.; Ren, H.-Y.; Li, D.-K. The Sugars and Organic Acids Composition in Fruits of Different Chinese Jujube Cultivars of Different Development Stages. *Acta Hortic. Sin.* **2016**, *43*, 1175–1185. [CrossRef]

33. Albertini, M.-V.; Carcouet, E.; Pailly, O.; Gambotti, C.; Luro, F.; Berti, L. Changes in organic acids and sugars during early stages of development of acidic and acidless citrus fruit. *J. Agric. Food Chem.* **2006**, *54*, 8335–8339. [CrossRef]
34. Bae, H.; Yun, S.K.; Jun, J.H.; Yoon, I.K.; Nam, E.Y.; Kwon, J.H. Assessment of organic acid and sugar composition in apricot, plumcot, plum, and peach during fruit development. *J. Appl. Bot. Food Qual.* **2014**, *87*, 24–29. [CrossRef]
35. Mao, Y.; Tian, S.; Qin, Y.; Cheng, S. An optimized organic acid human sensory sourness analysis method. *J. Sci. Food Agric.* **2021**, *101*, 5880–5887. [CrossRef]
36. Aslam, M.M.; Deng, L.; Wang, X.; Wang, Y.; Pan, L.; Liu, H.; Niu, L.; Lu, Z.; Cui, G.; Zeng, W.; et al. Expression patterns of genes involved in sugar metabolism and accumulation during peach fruit development and ripening. *Sci. Hortic.* **2019**, *257*, 108633. [CrossRef]
37. Zhang, W.; Lunn, J.E.; Feil, R.; Wang, Y.; Zhao, J.; Tao, H.; Guo, Y.; Zhao, Z. Trehalose 6-phosphate signal is closely related to sorbitol in apple (*Malus domestica Borkh.* cv. Gala). *Biol. Open* **2017**, *6*, 260–268. [CrossRef] [PubMed]
38. Wang, H.; Wang, J.; Mujumdar, A.S.; Jin, X.; Liu, Z.L.; Zhang, Y.; Xiao, H.W. Effects of postharvest ripening on physicochemical properties, microstructure, cell wall polysaccharides contents (pectin, hemicellulose, cellulose) and nanostructure of kiwifruit (*Actinidia deliciosa*). *Food Hydrocoll.* **2021**, *118*, 106808. [CrossRef]
39. Desnoues, E.; Gibon, Y.; Baldazzi, V.; Signoret, V.; Génard, M.; Quilot-Turion, B. Profiling sugar metabolism during fruit development in a peach progeny with different fructose-to-glucose ratios. *BMC Plant Biol.* **2014**, *14*, 336. [CrossRef] [PubMed]
40. Trad, M.; Boge, M.; Hamda, H.B.; Renard, C.M.G.C.; Harbi, M. The Glucose-Fructose ratio of wild Tunisian grapes. *Cogent Food Agric.* **2017**, *3*, 1374156. [CrossRef]
41. Song, J.; Bi, J.; Chen, Q.; Wu, X.; Lyu, Y.; Meng, X. Assessment of sugar content, fatty acids, free amino acids, and volatile profiles in jujube fruits at different ripening stages. *Food Chem.* **2019**, *270*, 344–352. [CrossRef]
42. Osorio, S.; Scossa, F.; Fernie, A.R. Molecular regulation of fruit ripening. *Front. Plant Sci.* **2013**, *14*, 198. [CrossRef] [PubMed]
43. Chervin, C. Should Starch Metabolism Be a Key Point of the Climacteric vs. Non-climacteric Fruit Definition? *Front. Plant Sci.* **2020**, *11*, 609189. [CrossRef]
44. Ikegaya, A.; Toyoizumi, T.; Ohba, S.; Nakajima, T.; Kawata, T.; Ito, S.; Arai, E. Effects of distribution of sugars and organic acids on the taste of strawberries. *Food Sci. Nutr.* **2019**, *7*, 2419–2426. [CrossRef]
45. Kader, A.A. Flavor quality of fruits and vegetables. *J. Sci. Food Agric.* **2008**, *88*, 1863–1868. [CrossRef]
46. Bacchichet, I.; Chiozzotto, R.; Bassi, D.; Gardana, C.; Cirilli, M.; Spinardi, A. Characterization of fruit quality traits for organic acids content and profile in a large peach germplasm collection. *Sci. Hortic.* **2021**, *278*, 109865. [CrossRef]
47. Nadeem, M.; Qureshi, T.M.; Ugulu, I.; Riaz, M.N.; An, Q.U.; Khan, Z.I.; Dogan, Y. Mineral, vitamin and phenolic contents and sugar profiles of some prominent date palm (*Phoenix dactylifera*) varieties of Pakistan. *Pak. J. Bot.* **2019**, *51*, 171–178. [CrossRef] [PubMed]
48. Czech, A.; Zarycka, E.; Yanovych, D.; Zasadna, Z.; Grzegorczyk, I.; Kłys, S. Mineral content of the pulp and peel of various citrus fruit cultivars. *Biol. Trace Elem. Res.* **2020**, *193*, 555–563. [CrossRef]
49. Aburto, N.J.; Hanson, S.; Gutierrez, H.; Hooper, L.; Elliott, P.; Cappuccio, F.P. Effect of increased potassium intake on cardiovascular risk factors and disease: Systematic review and meta-analyses. *BMJ* **2013**, *346*, f1378. [CrossRef]
50. DiNicolantonio, J.J.; Liu, J.; O'Keefe, J.H. Magnesium for the prevention and treatment of cardiovascular disease. *Open Heart* **2018**, *5*, e000775. [CrossRef]
51. WHO. *Potassium Intake for Adults and Children*; WHO: Geneva, Switzerland, 2012; pp. 1–52, ISBN 978-92-4-150482-9. Available online: https://www.who.int/publications/i/item/9789241504829 (accessed on 20 January 2023).
52. World Health Organization. *Vitamin and Mineral Requirements in Human Nutrition*; World Health Organization: Geneva, Switzerland, 2004. Available online: https://www.fao.org/3/y2809e/y2809e00.htm (accessed on 20 January 2023).

horticulturae

Article

The Main Physicochemical Characteristics and Nutrient Composition during Fruit Ripening of *Stauntonia obovatifoliola* Subsp. *Urophylla* (Lardizabalaceae)

Tianjiao Jia, Chen Feng, Shuaiyu Zou * and Puxin Gao *

Lushan Botanical Garden, Chinese Academy of Sciences, Jiujiang 332900, China
* Correspondence: zousy@lsbg.cn (S.Z.); gaopx@lsbg.cn (P.G.)

Abstract: *Stauntonia obovatifoliola* Hayata subsp. *urophylla* is a novel edible and healthy fruit in China, commonly known as "Jiuyuehuang" (September yellow). The fully ripe fruit of *S. obovatifoliola* subsp. *urophylla* has a soft fruit pulp texture, golden flesh, and sweet flavor which is very popular with the locals. In this paper, we have investigated the fruit appearance quality, physiochemical quality, and nutritional quality of *S. obovatifoliola* subsp. *urophylla* that was harvested at six stages (S1: 60 DAFB, S2: 90 DAFB, S3: 130 DAFB, S4: 160 DAFB, S5: 190 DAFB, S6: 205 DAFB). An increase in fruit size (including single fruit weight, fruit length, and fruit diameter) was related to the ripeness stage of fruit development. The total soluble solids, firmness, dry matter, sugar and starch showed remarkable changes as the fruit approached ripening (S5–S6 stage). The main sugar components in the fruit were fructose, glucose, and maltose. The contents of fructose, glucose, and total sugars in *S. obovatifoliola* subsp. *urophylla* fruit progressively increased from the S1 to the S6 stage while increasing sharply from the S4 to the S5 stage. As for the content of maltose and starch, they both showed an increasing trend from the S1 to the S4 stage but decreased sharply at the S5 stage. The vitamin B, vitamin C, total phenolics, total flavonoids, and amino acid levels showed an overall downward trend during fruit development. To our knowledge, this is the first study to compare the phytochemical characteristics, nutrient composition, and antioxidant content during the different fruit development stages. The results of this study may provide a scientific basis for clarifying the growth and development characteristics of *S. obovatifoliola* subsp. *urophylla* fruit and the further utilization of these excellent medicinal and edible germplasm resources.

Keywords: *Stauntonia obovatifoliola* subsp. *urophylla*; physicochemical; nutritional; ripening; maturity stage

Citation: Jia, T.; Feng, C.; Zou, S.; Gao, P. The Main Physicochemical Characteristics and Nutrient Composition during Fruit Ripening of *Stauntonia obovatifoliola* Subsp. *Urophylla* (Lardizabalaceae). *Horticulturae* **2023**, *9*, 29. https://doi.org/10.3390/horticulturae9010029

Academic Editors: Roberta Bulgari, Ada Baldi, Anna Lenzi and Antonios Chrysargyris

Received: 22 November 2022
Revised: 16 December 2022
Accepted: 21 December 2022
Published: 26 December 2022

1. Introduction

Stauntonia obovatifoliola Hayata subsp. *urophylla* (Hand.-Mazz.) H. N. Qin is a perennial woody liana. It belongs to the *Stauntonia* genus, has large and edible fruits, and is endemic to China [1]. The species is widely distributed in the Yangtze River basin provinces of China (Guangdong, Guangxi, Hunan, Jiangxi, Fujian, Zhejiang, etc.) (Figure 1), normally occurring at an altitude of 500–1500 m in forest edges, roadsides, or along streams in valleys [2]. Its fruits are usually called September yellow (Jiuyuehuang in Chinese) by local villagers, which is due to the fruit appearing yellow in color when ripe in Chinese lunar September [1]. The fully ripe fruit of *S. obovatifoliola* subsp. *urophylla* has a soft fruit pulp texture, golden flesh, and a sweet flavor, tasting like a mixture of persimmon and litchi. *S. obovatifoliola* subsp. *urophylla* fruits are rich in sugars, crude proteins, vitamins, amino acids, mineral elements, fat, total dietary fiber, and reducing sugars [3,4]. In addition to its edible value, *S. obovatifoliola* subsp. *urophylla* is also a traditional Chinese medicinal material. Its stems and fruits are rich in phenolics, flavonoids, triterpenes, sterides, and polysaccharides, and is usually used as an important medicinal material to treat rheumatic arthralgia, neuralgia, headache, heat strangury, trigeminal neuralgia, sciatica, and trauma pain [5–9]. These excellent nutritional and chemical compositions indicate that *S. obovatifoliola* subsp. *urophylla* is an alternative functional fruit that is worthy of development and utilization.

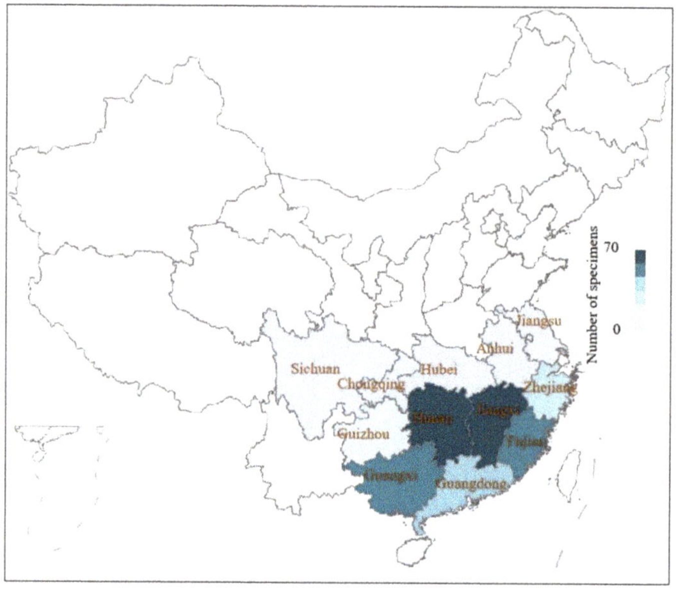

Figure 1. The geographic distribution of *Stauntonia obovatifoliola* subsp. *urophylla*. The distribution heat map was made based on the specimen data of *S. obovatifoliola* subsp. *urophylla* in a Chinese virtual herbarium.

In recent years, *S. obovatifoliola* subsp. *urophylla*, as a characteristic health fruit, has been widely cultivated in China. Even so, *S. obovatifoliola* subsp. *urophylla* is still an underexploited wild fruit in the infancy stage of domestication. Only a few studies on *S. obovatifoliola* subsp. *urophylla* breeding, germplasm evaluation and cultivation have been attempted, with even most efforts on phytochemical analyses [1,3,5,10,11]. Moreover, most of the reports about *S. obovatifoliola* subsp. *urophylla* were found on the local flora with only a brief description [12–16]. The lack of information on the change patterns of the fruit's physiological quality and physicochemical characteristics during fruit ripening impedes the development and utilization of *S. obovatifoliola* subsp. *urophylla*.

Fruit ripening is a very important physiological process in the later stage of fruit development which is usually accompanied by tremendous changes in the physical and chemical characteristics, such as sugar composition, fruit texture and color change, aromatic substances release, cell wall degradation, and so on, which ultimately affects the quality of fruit [17–20]. The analysis of these physical and chemical changes occurring during the development and ripening of fruit gives an insight into the underlying biochemical and physiological processes taking place [21] and is especially useful for determining fruit maturity. Although ripening is the process by which fruits attain their desirable color, flavor, nutritional quality, and textural properties, appropriate ripeness can effectively guarantee the commercial value of the fruit and prolong the storage time of the fruit. For instance, yellow peach will have a poor flavor if harvested too early, while the fruit can present a strong characteristic flavor when harvested late; however, if the olive fruit is harvested too late, the sugar content in the pulp will decrease and be prone to fibrosis [22,23]. In another example, in order to prolong the shelf life, winter jujube is usually harvested at the white maturity stage [24]. However, few studies are relevant to the change patterns in physicochemical indicators and nutrient composition of *S. obovatifoliola* subsp. *urophylla* at different fruit maturity stages. Therefore, it is necessary to explore the change patterns of *S. obovatifoliola* subsp. *urophylla* fruit's physiological parameters and physicochemical characteristics during fruit ripening, which provide basic physiological information for a better understanding of the ripening process of *S. obovatifoliola* subsp. *urophylla* fruit.

Since the optimum harvesting period is a considerable evaluation criterion for both food industries and consumers, the aim of the present study was to investigate the changes in the physicochemical characteristics, antioxidant content, and nutritional composition of *S. obovatifoliola* subsp. *urophylla* fruit at different maturity stages. The results of this work provide the basic dynamic change patterns of the fruit's physicochemical characteristics and explore a proper maturity stage at the harvest of *S. obovatifoliola* subsp. *urophylla* fruit with better quality, longer shelf life, and better market acceptability simultaneously.

2. Materials and Methods

2.1. Plant Materials

The plant materials were grown from seeds which were collected from Jiangxi Province, China, with relatively good comprehensive traits. A total of 56 plant materials were planted in 3 × 1.5 m in a fruit garden in Jiujiang City (29°38′ N, 115°59′ E), Jiangxi Province, China. The fruits of *S. obovatifoliola* subsp. *urophylla* free from insects, pests, and diseases were randomly harvested at six developmental stages (S1: 60 days after full bloom (DAFB), S2: 90 DAFB, S3: 130 DAFB, S4: 160 DAFB, S5: 190 DAFB, S6: 205 DAFB) until the fruits became fully ripe (S6 stage) in November (Figure 2). The *S. obovatifoliola* subsp. *urophylla* trees were four years old. Open pollination and a sprinkling irrigation system were used in the experimental farm. Due to a single tree of *S. obovatifoliola* subsp. *urophylla* being unable to provide enough fruit samples for all six stages, different trees were used at each stage in this experiment. In total, twenty fruit trees were used in this study, and ten fruits were harvested from each tree; thirty fruits were harvested at each stage. The fruits harvested from the same tree were considered as a sample at each stage. Three fruits' pulp was mixed into one biological replicate for sugars, vitamin B, vitamin C, total phenolics, total flavonoids, starch, and amino acid content analysis. We cut the fruit into slices with a knife and removed the peel and seeds to separate the pulp, and then the pulp was quickly treated with liquid nitrogen and stored at −80 °C until further analysis. In total, three biological replicates were conducted for each stage. The first sampling time was in June 2021, and the last sampling time was in November 2021. In this sampling period, the average daily temperature in Jiujiang was 20–28 °C, and the total precipitation was 471.80 mm.

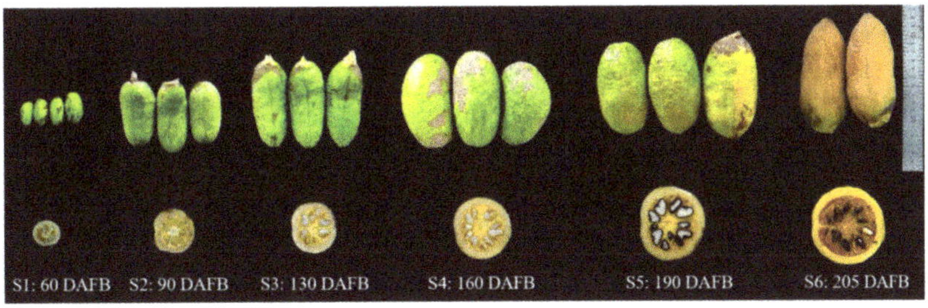

Figure 2. The fruits of *Stauntonia obovatifoliola* subsp. *urophylla* at different maturity stages.

2.2. Determination of Fruit Physical Parameters

The single fruit weight, fruit length, fruit diameter, and fruit shape index were evaluated immediately after the fruit samples were picked. The single fruit weight was recorded using an electronic balance. The fruit length and fruit diameter were measured using a digital caliper, and the ratio of fruit length to fruit diameter was the fruit shape index. The firmness of *S. obovatifoliola* subsp. *urophylla* fruit was determined by using a digital fruit firmness tester (GY-4, Zhejiang, China) equipped with a 3.5 mm cylinder probe, and the results were expressed as kg/cm^2. The dry matter content and moisture content in *S. obovatifoliola* subsp. *urophylla* fruit were measured by the drying–weighing method. Thin slices of fresh fruit were weighed using an electronic balance and then dried at 70 °C in an oven until their weights became constant. The dry matter content is calculated as the percentage of dry weight to fresh weight. Conversely, the moisture content is the percentage of lost weight to fresh weight.

2.3. Determination of Biochemical Parameters

The total soluble solids content (TSS) of *S. obovatifoliola* subsp. *urophylla* fruit was measured by an ATAGO (PAL-1) handheld digital refractometer, and the results were expressed in °Brix. The content of titratable acid (TA) was measured by the acid–base neutralization titration method, and the values were calculated as % of citric acid. The protein content in *S. obovatifoliola* subsp. *urophylla* fruit pulp was measured by the Kjeldahl method, and the results of protein content were expressed as N × 6.25 [25].

2.4. Determination of Carbohydrates

The sugar components (glucose, fructose, maltose, and sucrose) were measured by high-performance liquid chromatography (HPLC-E2695, Waters, Milford, MA, USA) equipped with a differential refraction detector, and the sample preparation and chromatographic conditions were conducted as described in the previous study [26].

The starch content of *S. obovatifoliola* subsp. *urophylla* fruit pulp was also determined by the method described in our previous study [26]. Briefly, approximately 1 g of fruit pulp samples and 20 mL of 80% ethanol were added together and vortex mixed to promote particle dispersion. Subsequently, the sample was incubated in the water bath at 80 °C for 30 min and stirred continuously until cooling. Then, the sample was filtered with 80% ethanol, and the filter residue was washed in a centrifuge tube by using hot distilled water and placed in a boiling water bath for gelatinization. Then, we added 2 mL of cold 9.2 mol/L perchloric acid and incubated for 15 min in a boiling water bath, and then filtered the mixture. The filtrate was poured into a 100 mL volumetric flask, and the filtrate residue was washed with distilled water. The absorbance was read at 490 nm by a differential refraction detector.

2.5. Determination of Vitamin B, Vitamin C, Total Phenolics, and Total Flavonoids

The Vitamin B_1, Vitamin B_2, Vitamin B_3, Vitamin B_6, and Vitamin C contents of *S. obovatifoliola* subsp. *urophylla* fruit pulp were measured by HPLC (HPLC-E2695, Waters, Milford, MA, USA) equipped with a diode array detector. Briefly, approximately 0.5 g of fruit pulp samples were weighed into a 15 mL centrifuge tube, and 5 mL of 0.1% hydrochloric acid aqueous solution was added. After 2 min of shock, the sample was extracted by ultrasonic wave (20 kHz, 30 °C) for 20 min and then centrifuged at a low temperature (4 °C) at a rate of 8000 r/min for 10 min. After the supernatant was filtered by a 0.45 μm filter membrane, then the sample was detected by HPLC. The detection conditions were as follows: a C18 chromatographic column (4.6 mm × 250 mm, 5 μm) was used for analysis; the mobile phase consisted of monopotassium phosphate and acetonitrile (80% and 20%, respectively) at a flow rate of 1.00 mL/min. The column temperature was 30 °C, and the injection volume was 10 μL.

The total phenolics content in the *S. obovatifoliola* subsp. *urophylla* fruit pulp was determined with Folin–Ciocalteu reagent following the method of Razzaq et al. [27]. Additionally, the total flavonoids content was measured following the method of Zhao et al. [28]. The detailed detection methods could be found in our previous study [26].

2.6. Determination of Amino Acids

The amino acid content in the *S. obovatifoliola* subsp. *urophylla* fruit pulp was measured by HPLC. Briefly, approximately 0.5 g of ground sample was put into the hydrolysis tube, 6 mL of 6 mol/L hydrochloric acid was added into the hydrolysis tube, and 3–4 drops of phenol were added, and then filled with high-purity nitrogen and sealed. The hydrolysis tube was placed in a constant temperature drying oven and hydrolyzed at 115 °C for 23 h. Then, we took out the hydrolysis tube from the oven and cooled it, adjusted the PH = 7 with sodium hydroxide solution, and added water to 7 mL for further analysis. Subsequently, 10 μL of extract solution was absorbed into the derivative tube, and 70 μL AccQ-Fluor borate buffer was added for vortex mixing. Then, another 20 μL AccQ-Fluor derivator was added to the tube, and vortex mixing was maintained for 10 s. We placed the tube at room temperature for one minute and then heated it in the oven at 55 °C for 10 min. Finally, we took out the hydrolysis tube from the oven and cooled it to ambient temperature, then transferred it to the fully recovered sample bottle for machine determination. AccQ·Tag column (3.9 × 150 mm) for amino acid analysis was used. The detection conditions were as follows, Mobile phase A: sodium acetate buffer solution; Mobile phase B: acetonitrile; Mobile phase C: pure water. Gradient elution was performed according to the Table 1.

Table 1. The elution gradient of mobile phase.

Time (min)	Flow Rate (mL/min)	Sodium Acetate Buffer (%)	Acetonitrile (%)	Pure Water (%)
start	1.0	100	0	0
0.5	1.0	98	2	0
13	1.0	95	5	0
19	1.0	91	9	0
29.5	1.0	83	17	0
33	1.0	0	60	40
36	1.0	100	0	0
45	1.0	100	0	0

The column temperature was 37 °C, and the injection volume was 10 μL. The excitation wavelength of the fluorescence detector was 250 nm, and the emission wavelength was 395 nm.

2.7. Statistical Analysis

SPSS v20.0 software was used to analyze the difference significance of the data (one-way ANOVA, LSD test, $p < 0.05$), and the results were expressed as means ± standard errors.

3. Results

3.1. Dynamic Changes of Appearance Quality during Fruit Development

The measurement results of the appearance quality of *S. obovatifoliola* subsp. *urophylla* fruit at different development stages were shown in Table 2. The single fruit weight began to increase rapidly at 60 DAFB, and the growth trend slowed down after 160 DAFB, entering the slow growth stage, and the single fruit weight increased to 192.72 ± 10.51 g during the period of the S1 to the S4 stages. As the fruit was close to ripening, the single fruit weight increased more slowly and had the highest weight value at the S6 stage. The growth patterns of the fruit length and fruit diameter were basically the same, and the rapid growth period of them was from 60 DAFB to 130 DAFB, and then they entered the slow growth period. The fruit shape index did not change much throughout the development period; the fruit length was about twice as long as the fruit diameter, indicating that the fruit length and fruit diameter developed simultaneously during the fruit growth.

Table 2. Dynamic changes in appearance quality during *S. obovatifoliola* subsp. *urophylla* fruit development.

Stages	Single Fruit Weight/g	Fruit Length/mm	Fruit Diameter/mm	Fruit Shape Index
S1	14.08 ± 1.28 e	47.79 ± 1.98 d	24.12 ± 0.58 d	1.98 ± 0.07 ab
S2	83.86 ± 4.45 d	84.24 ± 3.90 c	39.54 ± 0.87 c	2.14 ± 0.15 a
S3	133.75 ± 3.02 c	96.40 ± 3.95 b	52.54 ± 1.54 b	1.84 ± 0.08 b
S4	192.72 ± 10.51 b	111.62 ± 5.03 a	53.94 ± 1.65 b	2.08 ± 0.12 ab
S5	230.57 ± 8.15 a	114.21 ± 3.94 a	59.57 ± 0.24 a	1.92 ± 0.07 ab
S6	240.40 ± 2.22 a	118.77 ± 3.27 a	62.86 ± 1.93 a	1.89 ± 0.03 ab

Means with different letters within the same column indicate statistical differences at the $p < 0.05$ level. S1: 60 DAFB (days after full bloom), S2: 90 DAFB, S3: 130 DAFB, S4: 160 DAFB, S5: 190 DAFB, S6: 205 DAFB.

3.2. Dynamic Changes of Total Soluble Solids, Titratable Acidity, Dry Matter, and Fruit Firmness during Fruit Development

The total soluble solids content fluctuated below 2° Brix in the early stages (S1–S4). Whereas there were no significant differences between those stages, thereafter, the total soluble solids content of the fruit pulp increased sharply during the S5 and S6 stages and reached the maximum value of 13.52° Brix in the ripening stage S6 (Figure 3A). Although the titratable acidity content maintained an increasing trend during the whole fruit development period, the titratable acidity content of the *S. obovatifoliola* subsp. *urophylla* fruit maintained at a low level, and the values fluctuated in the range of 0.62–0.84% during the period of the S1 to the S6 maturity stage (Figure 3B).

The content of dry matter content increased continuously in the early stage of fruit development (S1–S4) and reached the maximum value of 27.90% at the S4 stage but decreased slightly during the period from S5 to S6 (Figure 3C).

The fruit firmness of *S. obovatifoliola* subsp. *urophylla* was significantly ($p < 0.05$) affected by the fruit development stage. The fruit firmness also maintained an increasing trend during the early fruit development period and reached the maximum value of 127.53 kg/cm^2 at the S4 stage but decreased sharply during the period from S5 to S6, and had a minimum value of 15.75 kg/cm^2 at the S6 stage (Figure 3D).

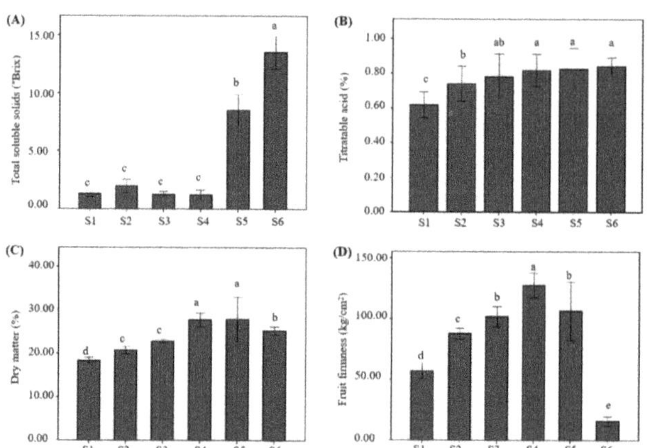

Figure 3. Total soluble solids (**A**), titratable acidity (**B**), dry matter (**C**), and fruit firmness (**D**) in *S. obovatifoliola* subsp. *urophylla* fruit at different ripening stages. Means with different letters within the same fruit character indicate statistical differences at the $p < 0.05$ level using the LSD test. S1: 60 DAFB (days after full bloom), S2: 90 DAFB, S3: 130 DAFB, S4: 160 DAFB, S5: 190 DAFB, S6: 205 DAFB.

3.3. Dynamic Changes of Carbohydrate Contents during Fruit Development

Carbohydrate contents in *S. obovatifoliola* subsp. *urophylla* fruit showed significant changes during fruit development and ripening (Figure 4). Only three (fructose, glucose, and maltose) of the four tested soluble sugars were detected in the fruit pulp of *S. obovatifoliola* subsp. *urophylla*, of which sucrose was not detected by HPLC. The fructose contents were very low before the S5 stage and fluctuated from 0.10 g/100 g FW to 0.33 g/100 g FW. As Figure 4A showed, the fructose content accumulated rapidly during the S5 and S6 stages and reached the maximum value of 5.35 g/100 g FW at the mature stage (S6). Similarly, the glucose showed the same accumulation pattern as the fructose, mainly accumulated during the S5 and S6 stages and reaching the maximum value of 7.02 g/100 g FW at the S6 stage (Figure 4B). The maltose content of *S. obovatifoliola* subsp. *urophylla* fruit continuously accumulated from the S1 to the S3 stage (0.72–1.22 g/100 g FW) and reached the maximum value at the S3 stage, although it had a slight decrease at the S4 stage, whereafter, a sharp and significant decline during the S5 and S6 stages were detected (0.16 g/100 g FW and 0.08 g/100 g FW, respectively) (Figure 4C). The contents of the total sugars of the *S. obovatifoliola* subsp. *urophylla* fruit were relatively low and changed little in the early stages of fruit development (S1–S4) and began to increase significantly from the S5 stage and reached the maximum value at the S6 stage (12.45 g/100 g FW) (Figure 4D). The starch content of *S. obovatifoliola* subsp. *urophylla* fruit continuously accumulated from the S1 to the S4 stage (0.21–1.09 g/100 g FW) and reached the maximum value at the S4 stage, but had a sharp decrease at the S5 stage, and finally reached the minimum value at the mature stage (S6) (0.07 g/100 g FW) (Figure 4E).

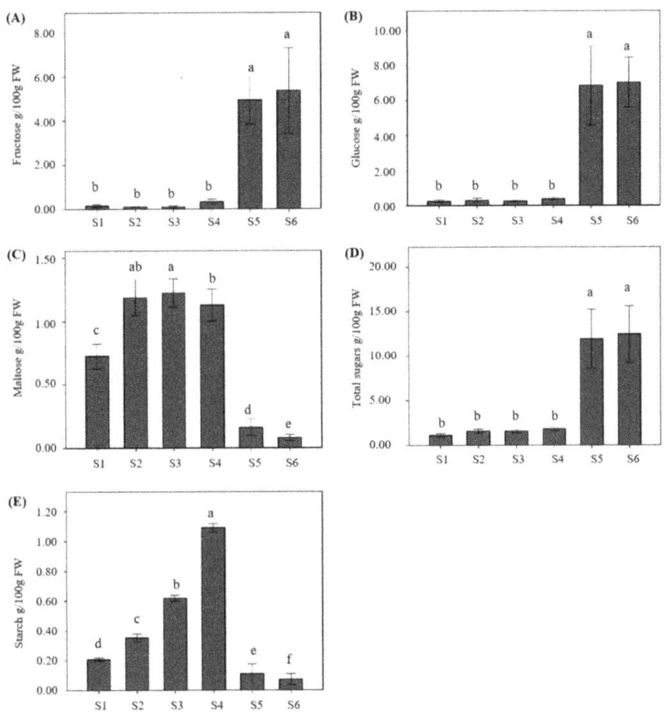

Figure 4. Dynamic changes of concentrations of fructose (**A**), glucose (**B**), maltose (**C**), total sugars (the sum of glucose, fructose, and maltose) (**D**), and starch (**E**) during the fruit development of *S. obovatifoliola* subsp. *urophylla*. The different letters in each figure indicate significant differences at the $p < 0.05$ level using the LSD test. S1: 60 DAFB (days after full bloom), S2: 90 DAFB, S3: 130 DAFB, S4: 160 DAFB, S5: 190 DAFB, S6: 205 DAFB.

3.4. Dynamic Changes of Vitamin B, Vitamin C, Total Phenolics, Total Flavonoids, and Protein Contents during Fruit Development

The changes in antioxidant component and protein contents during *S. obovatifoliola* subsp. *urophylla* fruit development were shown in Table 3. The vitamin B1 content increased from the S1 to the S2 stage and reached the maximum value at the S2 stage (20.24 ± 1.68 mg/100 g FW); thereafter, the content of vitamin B1 continued to decline from the S2 to the S6 stage; particularly, between the S5 and S6 stages, a sharp and significant decline has been observed. The content of vitamin B2 has the maximum value at the S1 stage (191.21 ± 2.99 mg/100 g FW), then fluctuated between 31.08 ± 1.27 mg/100 g FW and 38.65 ± 1.57 mg/100 g FW; subsequently, the vitamin B2 content declined slightly and tended to be stable during the S5 and S6 stages. The vitamin B3 content also has the maximum value at the S1 stage (10.53 ± 0.71 mg/100 g FW), then fluctuated between 3.08 ± 0.16 mg/100 g FW and 4.68 ± 0.28 mg/100 g FW, and finally reached the minimum value at the S6 stage (2.03 ± 0.30 mg/100 g FW). As displayed in Table 3, the content of vitamin B6 at the six maturity stages declined continuously with the delaying of the fruit's development, and finally reached the minimum value at the mature stage (S6) (0.13 ± 0.01 mg/100 g FW). The content of vitamin C indicated the same accumulation pattern as vitamin B1. That is, the content of vitamin C increased from the S1 to the S2 stage and reached the maximum value at the S2 stage (608.58 ± 7.28 mg/100 g FW). Thereafter, the content of vitamin C continued to decline from the S2 to the S6 stage; particularly, between the S4 and S5 stages, a sharp and significant decline was observed, and it finally tended to be stable during the S5 and S6 stages.

As shown in Table 3, the protein content continuously declined during the whole development period and reached the minimum value at the S6 stage (0.48 ± 0.00 g/100 g FW). The content of total phenolics declined from the S1 to the S2 stage, then increased at the S3 stage, and declined again at the S4 stage, whereas a significant rise was found from the S4 to the S5 stage; thereafter, there was no significant difference between the S5 and S6 stages. Similarly, the total flavonoids indicated the same accumulation pattern as the total phenolics in the early stage of fruit development (S1–S3) and then showed a continuing downward trend until reaching the minimum value at the mature stage (13.85 ± 0.80 mg/100 g FW).

Table 3. Dynamic changes of antioxidant component and protein content during *S. obovatifoliola* subsp. *urophylla* fruit development.

Stages	Vitamin B_1 mg/100 g FW	Vitamin B_2 mg/100 g FW	Vitamin B_3 mg/100 g FW	Vitamin B_6 mg/100 g FW	Vitamin C mg/100 g FW	Protein Content g/100 g	Total Phenolics mg/100 g FW	Total Flavonoids mg/100 g FW
S1	13.88 ± 0.66 b	191.21 ± 2.99 a	10.53 ± 0.71 a	1.39 ± 0.08 a	553.88 ± 18.12 b	2.08 ± 0.04 a	1491.92 ± 10.48 a	622.92 ± 0.47 a
S2	20.24 ± 1.68 a	31.08 ± 1.27 c	4.12 ± 0.32 bc	0.62 ± 0.03 b	608.58 ± 7.28 a	1.54 ± 0.01 b	295.40 ± 3.65 d	25.20 ± 0.46 d
S3	18.62 ± 0.47 a	33.77 ± 1.07 bc	3.08 ± 0.16 cd	0.51 ± 0.03 b	464.12 ± 13.49 c	1.39 ± 0.00 c	376.91 ± 1.80 c	73.09 ± 0.28 b
S4	14.87 ± 1.21 b	38.65 ± 1.57 b	4.68 ± 0.28 b	0.58 ± 0.03 b	332.62 ± 6.52 d	1.35 ± 0.04 c	323.39 ± 5.19 d	39.08 ± 0.55 c
S5	14.53 ± 1.17 b	18.17 ± 0.97 d	4.66 ± 0.75 b	0.18 ± 0.03 c	1.02 ± 0.04 e	0.93 ± 0.08 d	531.10 ± 20.23 b	15.09 ± 0.93 e
S6	2.58 ± 0.21 c	22.38 ± 2.47 d	2.03 ± 0.30 d	0.13 ± 0.01 c	1.22 ± 0.19 e	0.48 ± 0.00 e	512.38 ± 9.62 b	13.85 ± 0.80 e

Means with different letters within the same column indicate statistical differences at the $p < 0.05$ level. S1: 60 DAFB (days after full bloom), S2: 90 DAFB, S3: 130 DAFB, S4: 160 DAFB, S5: 190 DAFB, S6: 205 DAFB.

3.5. Dynamic Changes of Amino Acid Composition during Fruit Development

We detected the content of 17 common free amino acids in the fruit pulp of *S. obovatifoliola* subsp. *urophylla* at different development stages, and the results were shown in Table 4. In general, the content of amino acids at the six development stages declined continuously with the fruit development and has the minimum value at the mature stage, including glutamic acid (Glu), glycine (Gly), alanine (Ala), tyrosine (Tyr), valine (Val), methionine (Met), lysine (Lys), isoleucine (Ile), leucine (Leu), proline (Pro), and phenylalanine (Phe). However, some amino acids showed a spike and then declined in an overall decline trend, such as aspartic acid (Asp) showing a spike at the S2 stage, threonine (Thr) showing a spike at the S4 stage, and cysteine (Cys) showing a spike at the S5 stage, whereas the content of serine (Ser), histidine (His), and arginine (Arg) have maximum values in the S1 stage and showed a spike at the S5 stage. The content of total amino acids (TAAs) declined continuously with the fruit development and has the minimum value at the mature stage (336.89 ± 11.98 mg/100 g). Similarly, essential amino acids (EAAs) indicated the same accumulation pattern as the TAAs. In particular, the proportion of EAAs, including Thr, Val, Met, Lys, Ile, Leu, and Phe, has a maximum value at the mature stage (38.94%).

Table 4. Dynamic changes of amino acid composition during *S. obovatifoliola* subsp. *urophylla* fruit development.

Amino Acids (mg/100 g)	Fruit Development Stages					
	S1	S2	S3	S4	S5	S6
Aspartic acid (Asp)	324.90 ± 4.88 b	433.71 ± 4.40 a	393.44 ± 5.56 a	332.24 ± 8.34 b	162.33 ± 32.59 c	33.45 ± 1.15 d
Serine (Ser)	148.56 ± 1.63 a	75.05 ± 2.08 c	75.91 ± 0.36 c	78.36 ± 0.82 c	86.12 ± 2.35 b	31.30 ± 1.21 d
Glutamic acid (Glu)	103.97 ± 3.37 a	50.13 ± 0.77 b	44.46 ± 0.50 b	44.58 ± 0.45 b	32.20 ± 4.29 c	23.89 ± 1.57 d
Glycine (Gly)	67.18 ± 0.71 a	33.29 ± 0.34 b	28.81 ± 1.15 c	31.92 ± 0.63 bc	23.91 ± 2.51 d	17.53 ± 0.9 e
Threonine (Thr) *	71.02 ± 1.25 a	34.38 ± 0.70 b	30.31 ± 0.92 c	34.50 ± 0.48 b	15.01 ± 0.72 d	17.01 ± 0.78 d
Histidine (His)	38.18 ± 0.76 a	21.05 ± 0.95 c	17.18 ± 0.43 cd	20.25 ± 0.71 c	27.77 ± 3.71 b	12.72 ± 0.29 d
Arginine (Arg)	141.98 ± 2.05 a	69.54 ± 0.90 c	65.17 ± 0.60 c	66.58 ± 1.25 c	85.75 ± 2.66 b	41.31 ± 1.86 d
Alanine (Ala)	40.21 ± 0.78 a	19.47 ± 0.45 b	18.48 ± 0.42 b	18.83 ± 0.20 b	17.92 ± 1.10 b	9.38 ± 0.31 c
Proline (Pro)	47.07 ± 14.05 a	33.62 ± 1.49 ab	30.17 ± 0.94 ab	29.49 ± 0.83 ab	34.73 ± 1.49 ab	15.35 ± 0.57 b
Cysteine (Cys)	6.35 ± 0.27 b	4.31 ± 0.18 bcd	4.58 ± 0.25 bc	2.81 ± 0.16 cd	27.43 ± 1.80 a	1.86 ± 0.18 d
Tyrosine (Tyr)	55.03 ± 1.35 a	26.55 ± 1.60 b	25.94 ± 0.68 b	26.80 ± 0.63 b	24.72 ± 0.95 b	18.90 ± 0.72 c
Valine (Val) *	87.50 ± 1.45 a	42.78 ± 1.16 b	36.90 ± 1.37 c	39.51 ± 1.12 bc	27.37 ± 2.36 d	20.30 ± 0.71 e
Methionine (Met) *	13.41 ± 0.49 a	5.88 ± 0.36 bc	6.83 ± 0.17 b	4.82 ± 0.11 cd	5.44 ± 0.54 cd	4.31 ± 0.27 d
Lysine (Lys) *	85.47 ± 1.70 a	40.02 ± 2.07 b	37.60 ± 0.93 b	38.10 ± 0.97 b	27.42 ± 2.60 c	20.82 ± 0.61 d
Isoleucine (Ile) *	112.80 ± 3.08 a	55.06 ± 1.54 b	52.58 ± 1.15 b	51.39 ± 1.21 b	37.46 ± 3.67 c	27.60 ± 0.91 d
Leucine (Leu) *	98.02 ± 4.13 a	49.33 ± 1.65 b	44.96 ± 2.14 b	47.07 ± 1.80 b	32.64 ± 3.49 c	24.97 ± 0.69 c
Phenylalanine (Phe) *	64.15 ± 1.25 a	31.12 ± 0.94 b	27.75 ± 0.56 b	28.40 ± 0.82 b	20.89 ± 2.22 c	16.17 ± 0.49 d
Total amino acids (TAAs)	1505.81 ± 25.32 a	1025.28 ± 14.88 b	941.07 ± 13.59 bc	895.65 ± 17.42 c	689.08 ± 59.57 d	336.89 ± 11.98 e
Essential amino acids (EAAs)	532.37 ± 10.03 a 35.35%	258.58 ± 7.74 b 25.22%	236.93 ± 4.58 b 25.18%	243.78 ± 5.54 b 27.22%	166.21 ± 15.54 c 24.12%	131.20 ± 4.26 d 38.94%

Means with different letters within the same line indicate statistical differences at the $p < 0.05$ level. * indicates essential amino acids, and the percentage number (%) represents the proportion of essential amino acids among the total amino acids. S1: 60 DAFB (days after full bloom), S2: 90 DAFB, S3: 130 DAFB, S4: 160 DAFB, S5: 190 DAFB, S6: 205 DAFB.

4. Discussion

Stauntonia obovatifoliola subsp. *urophylla* is a novel edible and healthy fruit and has tremendous potential for exploitation and utilization. In this study, the dynamic changes in the fruit quality, sugar composition and content, antioxidant component and content, and amino acids content of *S. obovatifoliola* subsp. *urophylla* during fruit development were detected and analyzed.

The detection results of the fruit's appearance quality showed that the fruit increased rapidly in the early stages of development. The single fruit weight of *S. obovatifoliola* subsp. *urophylla* at the S2, S3, S4, S5, and S6 stages increased by 495.60%, 59.49%, 44.09%, 19.64%, and 4.26%, respectively. For the fruit length, the growth rates at each stage were 76.27%, 14.43%, 15.79%, 2.32%, and 3.99%, respectively. The highest growth rate of fruit diameter also occurred at the S2 stage (63.93%), and the lowest growth rate was found at the S4 stage (2.66%), whereafter, the growth rate showed a spike at the S5 stage (S5–S6, 10.44%, and 5.52%, respectively). The results of the fruit shape index showed that the fruit shape index of *S. obovatifoliola* subsp. *urophylla* was basically stable between 1.8–2.2 during the fruit development period, and the difference between each stage was not obvious. In particular, the values of fruit weight, fruit length, and fruit diameter at the mature stage were much higher than those that grew in the wild [3]. Moreover, most of the wild *S. obovatifoliola* subsp. *urophylla* fruits are small and have a low edible ratio, low stress resistance, and poor appearance quality. Fortunately, the wide range of geographical distribution of *S. obovatifoliola* subsp. *urophylla* provides substantial genetic diversity and rich wild germplasm resources for breeders to select superior genotypes with excellent comprehensive characteristics through resource exploration and evaluation.

Total soluble solids and titratable acid are the main components of fruit flavor quality and nutritional composition which were also considered as crucial parameters of fruit ripening [29]. Additionally, the importance of detecting the change of TSS, TA, and TSS/TA has also been demonstrated by many studies in different fruits, such as strawberry, sweet cherry, orange, mulberry, etc. [30–32]. In this experiment, the TSS content of the *S. obovatifoliola* subsp. *urophylla* fruit showed an increasing trend during the whole fruit development period. Exactly, the TSS content kept at low levels during the early fruit development period (S1–S4), while the TSS content accumulated sharply during the S5–S6 stages, which was mainly due to the breakdown of starch and the accumulation of sugars. Meanwhile, the TA content of the *S. obovatifoliola* subsp. *urophylla* fruit remained at low levels during the fruit ripening process, which is consistent with previous reports on *S. obovatifoliola* subsp. *urophylla* fruit [3]. Thus, the high ratio of TSS to TA resulted in the sweet flavor of *S. obovatifoliola* subsp. *urophylla* fruit. Moreover, the TSS content changed sharply when the fruit neared ripening, which could be considered as an alternative maturity indicator of *S. obovatifoliola* subsp. *urophylla* fruit.

The dry matter content not only is a significant parameter to evaluate the carbon incorporation at different development stages of fruits but is also an important indicator of fruit flavor quality and texture [33,34]. The dry matter content of the *S. obovatifoliola* subsp. *urophylla* fruit continuously increased in the early stages and then declined significantly at the mature stage, which was mainly due to the direct relation between an increase in fruit size and the degradation of starch. Fruit firmness usually has a significant impact on fruits' market value, consumer acceptance, and shelf life [26,35]. The firmness declined sharply when the fruit of *S. obovatifoliola* subsp. *urophylla* neared maturity and softening. Unlike *Akebia* fruits (a relative genus of *Stauntonia*), the *S. obovatifoliola* subsp. *urophylla* fruit did not crack when fully ripe; thus, it could be picked after full maturity (S6 stage), and it tasted better as a fresh fruit. However, considering the long-distance transportation, we suggest that the best time point to harvest the fruit may be at the S5 stage as the fruit is maintained at a suitable hardness that could bear long-distance transportation.

Sugars are not only important components of fruit flavor and nutritional composition but also play important signaling factor roles during plant growth and are also involved in regulating gene expression during plant development [36,37]. The major sugars identified in *S. obovatifoliola* subsp. *urophylla* fruit pulp are fructose, glucose, and maltose. The contents of fructose, glucose, and total sugars showed the same accumulation trend, which firstly increased slightly in the early fruit development periods (S1–S4), thereafter increased sharply at the S5 stage, and there were no significant dynamic changes at the S5 and S6 stages. The changing trends of maltose and starch during the fruit's development and ripening were basically similar; both continuously accumulated during the early fruit development periods but declined sharply as the fruit approached ripening. This opposite change trend of monosaccharides and polysaccharides was mainly due to the hydrolysis of carbohydrates as the fruit ripened, similar to strawberry, sweet cherry, bananas, etc. [30,38,39].

Vitamins, apart from being an important nutrient in daily diet, are also potent antioxidant components [40–42]. The results of this study showed that the fruit of *S. obovatifoliola* subsp. *urophylla* was rich in vitamin B (B_1, B_2, B_3, B_6), vitamin C, phenolics, and flavonoids. In particular, the content of vitamin B_2 was the most abundant vitamin B at the mature stage, while the vitamin C content at the mature stage was in accordance with a previous report [6]. The vitamin content of *S. obovatifoliola* subsp. *urophylla* fruit showed an overall downward trend during fruit development, but there was a large fluctuation in this process, especially the vitamin C content, which decreased sharply when the fruit was near ripening (S5–S6). Phenolics and flavonoids compounds are important secondary metabolites of plants, which can help plants' resistance to bacteria and protect cells from oxidative damage [43,44]. In this study, the content of phenolics declined from the S1 to the S2 stage, followed by a subsequent increase in fluctuation. The flavonoids content declined firstly from the S1 to the S2 stage and increased at the S3 stage, followed by a subsequent sustained decrease in further fruit development in *S. obovatifoliola* subsp. *urophylla* fruit.

The dynamic change of antioxidant content was closely related to fruit development and ripening; the content of antioxidants could also be influenced by structural genes, temperature, light intensity, etc. [45,46]. The general decrease in the content of vitamins, phenolics, and flavonoids is mainly due to the oxidation of oxidase during fruit maturity or the effect of dilution as fruit increases in size [47,48].

Amino acids are important nutrient elements for most of life and also play some key roles in various biological reactions, such as signaling pathways, ATP generation, redox balance, nucleotide synthesis, cellular immunity, etc. [49–53]. In this study, 17 common free amino acids and 7 essential amino acids were detected by HPLC. The content of amino acids showed an overall downward trend during fruit development, which might be related to the conversion and expenditure of amino acids. Another reason may be the dilution effect of increasing fruit size. The abundance of amino acids in the fruit of *S. obovatifoliola* subsp. *urophylla* documented its excellent medicinal and edible value.

Compared to some common fruits, *S. obovatifoliola* subsp. *urophylla* has more advantages in certain nutrients (Table 5). For example, the total soluble solids, total sugars, fructose, and glucose are higher than apple, peach, kiwifruit, and strawberry, indicating that the *S. obovatifoliola* subsp. *urophylla* fruit has a sweeter flavor than some common fruits. *S. obovatifoliola* subsp. *urophylla* fruit has more protein than apple and cherry but less than banana, grape, peach, kiwifruit, and strawberry. As for the content of total amino acids, *S. obovatifoliola* subsp. *urophylla* is on par with grape, kiwifruit, strawberry, and cherry whereas less than apple, banana, and peach. Surprisingly, the contents of vitamin B_1, vitamin B_2, and vitamin B_3 of *S. obovatifoliola* subsp. *urophylla* are much higher than those common fruits, whereas it has less vitamin C than those common fruits. It should be pointed out that the determination of the nutritional composition of *S. obovatifoliola* subsp. *urophylla* in this paper is not comprehensive, but it can still prove that *S. obovatifoliola* subsp. *urophylla* is worthy of being exploited and utilized as a medicinal and edible fruit crop.

Table 5. The nutrient composition of *S. obovatifoliola* subsp. *urophylla* fruit at the ripening stage [54,55].

Parameter	*S. obovatifoliola* Subsp. *Urophylla*	Apple	Banana	Grape	Peach	Kiwifruit	Strawberry	Cherry
Total soluble solids (%)	13.52	12.00	22.63	17.51	7.82	7.58	7.08	14.30
total sugars (g/100 g)	12.45	11.07	19.55	15.30	6.17	6.99	5.93	11.90
fructose (g/100 g)	5.35	4.66	7.96	6.02	0.89	1.96	1.74	7.63
glucose (g/100 g)	7.02	1.39	2.41	7.56	0.75	2.38	2.42	1.44
protein (g/100 g)	0.48	0.41	1.26	0.55	0.78	1.15	0.90	0.32
Total amino acids (mg/100 g)	689.08	2706.92	7623.91	368.33	4580.00	629.27	667.00	573.00
vitamin B_1 (mg/100 g)	2.58	0.02	0.02	0.03	0.01	0.05	0.02	0.02
vitamin B_2 (mg/100 g)	22.38	0.02	0.04	0.02	0.02	0.02	0.02	0.02
vitamin B_3 (mg/100 g)	2.03	0.20	0.70	0.25	0.30	0.30	0.30	0.60
vitamin C (mg/100 g)	1.22	2.84	8.00	4.79	5.17	89.89	74.80	11.12

5. Conclusions

In conclusion, the dynamic changes in the fruit's appearance, physiochemical quality, and nutritional quality during the development of *S. obovatifoliola* subsp. *urophylla* were detected and analyzed. This study suggested that the maturity stage had a significant effect on nutritional properties and physiochemical parameters during *S. obovatifoliola* subsp. *urophylla* fruit ripening and softening. Particularly, the values of TSS, firmness, dry matter, fructose, glucose, maltose, starch, vitamin B, vitamin C, total phenolics, and total flavonoids of *S. obovatifoliola* subsp. *urophylla* fruit showed significant changes during the transition to physiological maturity. In view of the nutrient content and pulp texture, the fruit of *S. obovatifoliola* subsp. *urophylla* picked at the S5 maturity stage was more suitable for long-distance transportation. The results of this study laid a foundation for clarifying the growth and development characteristics of *S. obovatifoliola* subsp. *urophylla* fruit and the further utilization of these excellent medicinal and edible germplasm resources.

Author Contributions: Writing—original draft preparation, T.J.; investigation, C.F.; conceptualization, writing—review and editing, S.Z.; funding acquisition, P.G. All authors have read and agreed to the published version of the manuscript.

Funding: This research was funded by the Plant Germplasm Innovation Program of Chinese Academy of Sciences (KFJ-BRP-007-001) and Youth Fund of Lushan Botanical Garden, the Chinese Academy of Sciences (2021 ZWZX07).

Institutional Review Board Statement: Not applicable.

Informed Consent Statement: Not applicable.

Data Availability Statement: Not applicable.

Conflicts of Interest: The authors declare no conflict of interest.

References

1. Zou, S.; Yao, X.; Zhong, C.; Gao, P.; Wang, Z.; Huang, H. Phenotypic characterization of *Stauntonia obovatifoliola* Hayata subsp. urophylla germplasm: A potential new fruit crop. *Genet. Resour. Crop Evol.* **2020**, *67*, 1037–1050.
2. Editorial Commission of Chinese Flora of Chinese Academy of Sciences. *Flora of China*; Science Press: Beijing, China, 2001.
3. Wang, Y.; Xing, W.; He, X.; Duan, W.; Huang, W.; Gong, C. Differential comparison among different geographical provenances of *Stauntonia urophylla*. *J. Cent. South Univ. For. Technol.* **2015**, *35*, 55–60.
4. Li, R.; Ru, Y.; Feng, L.; Wang, Z.; He, X.; Zhang, X. A comparative study of nutrient composition, bioactive properties and phytochemical characteristics of *Stauntonia obovatifoliola* flesh and pericarp. *Front. Nutr.* **2022**, *9*, 1013971. [CrossRef] [PubMed]
5. Hong, Z.M.; Yang, T.M. Chemical base of Yao's medication principle. *J. South-Cent. Univ. Natl.* **2010**, *29*, 37–40.
6. Peng, X.B.; Gao, W.L.; Hu, D.Q.; Ma, F.F.; Fu, L.G.; Deng, Q.; Wei, Y. Chemical constituents from the aerial part of *Stauntonia obovatifoliola* Hayata subsp. urophylla. *J. Chin. Med. Mater.* **2013**, *36*, 1795–1798.
7. Lu, X.R.; Qiu, F.; Pan, X.Q.; Li, J.; Wang, M.Y.; Gong, M.X. Simultaneous quantitative analysis of nine triterpenoid saponins for the quality control of *Stauntonia obovatifoliola* Hayata subsp. intermedia stems. *J. Sep. Sci.* **2014**, *37*, 3632–3640. [CrossRef]
8. Lu, X.R.; Wang, X.M.; Wang, Z.M.; Chen, X.Q.; Wang, M.Y.; Gong, M.X. Triterpenoid Saponins from *Stauntonia obovatifoliola* Hayata ssp. urophylla. *Helv. Chim. Acta* **2015**, *98*, 245–252. [CrossRef]
9. Thomas, S.S.; Cha, Y.S.; Kim, K.A. Perilla oil alleviates high-fat diet-induced inflammation in the colon of mice by suppressing nuclear factor-kappa b activation. *J. Med. Food* **2020**, *23*, 818–826. [CrossRef]
10. Chen, X.D.; Tian, W.; Deng, R.H.; Liu, Y.; Deng, H.Y. Chemical constituents from *Stauntonia obovatifoliola* subsp. Urophylla. *Chin. Tradit. Herb. Drugs* **2013**, *44*, 671–673.
11. Zhou, W.J.; Chen, Q.P.; Zeng, Y.B.; Wang, G.X. Content Determination of Oleanolic Acid in *Stauntonia obovatifoliola* Hayata subsp. urophylla by RP-HPLC. *Chin. J. Inf. TCM* **2015**, *22*, 92–94.
12. Shan, Z.; Mu, Z.; Qin, H. Study on the species identification and distribution of Lardizabalaceae medicinal plants in Jiangxi province. *China Med. Pharm.* **2018**, *8*, 53–57.
13. Wang, Y.; He, X.; Li, J.; Gong, C.; Zhou, W. Evaluation of adaptability on germplasm resources of Lardizabalaceae. *South China For. Sci.* **2018**, *46*, 24–27.
14. Wu, L.; Xu, W.; He, S. Species and distributuin of medicinal plants of Lardizabalaceae in China. *J. Anhui Agric. Sci.* **2010**, *38*, 14325–14326.
15. You, C.; Huang, H.; Mei, Y. Study of Lardizabalaceae resources and cultivation in southern Jiangxi Province. *South China For. Sci.* **2018**, *46*, 22–24.
16. Zhong, W.; Cao, L.; Zhong, W.; Mu, Z.; Du, X.; Zhong, G. Survey resources of medicinal plants of Lardizabalaceae in Jiangxi Province. *J. Jiangxi Univ. TCM* **2016**, *28*, 63–64.
17. Lelièvre, J.M.; Latchè, A.; Jones, B.; Bouzayen, M.; Pech, J.C. Ethylene and fruit ripening. *Physiol. Plant.* **1997**, *101*, 727–739. [CrossRef]
18. Palma, J.M.; Corpas, F.J.; del Rio, L.A. Proteomics as an approach to the understanding of the molecular physiology of fruit development and ripening. *J. Proteomics* **2011**, *74*, 1230–1243. [CrossRef]
19. Liu, R.; How-Kit, A.; Stammitti, L.; Teyssier, E.; Rolin, D.; Mortain-Bertrand, A.; Halle, S.; Liu, M.; Kong, J.; Wu, C.; et al. A DEMETER-like DNA demethylase governs tomato fruit ripening. *Proc. Natl. Acad. Sci. USA* **2015**, *112*, 10804–10809. [CrossRef]
20. Shinozaki, Y.; Nicolas, P.; Fernandez-Pozo, N.; Ma, Q.; Evanich, D.J.; Shi, Y.; Xu, Y.; Zheng, Y.; Snyder, S.I.; Martin, L.B.B.; et al. High-resolution spatiotemporal transcriptome mapping of tomato fruit development and ripening. *Nat. Commun.* **2018**, *9*, 364. [CrossRef]
21. Dabesor, P.A.; Sanni, D.M.; Kolawole, A.O.; Enujiugha, V.N.; Lawal, O.T.; Edeh, A.T. Changes in physicochemical properties and enzymes associated with ripening of snake tomato (*Trichosanthes Cucumerina* L.) fruit. *Biocatal. Agric. Biotechnol.* **2022**, *40*, 102313. [CrossRef]
22. Yu, Y.; Zhang, X.; Zhang, L.P.; Gao, Z.S.; Jia, H.J. Effect of both harvest maturities and storage temperatures on fruit quality of'Jinxiu' yellow peach. *Sci. Technol. Food Ind.* **2015**, *36*, 334–338.

23. Kong, X.J.; Lin, H.T.; Zhou, H.; Lin, Y.F.; Chen, Y.H.; Wang, H. Study on the optimum harvesting date of fresh-eating Chinese olive fruit and its quality assessment parameters. *Storage Process* **2016**, *16*, 6–14.
24. Zozio, S.; Servent, A.; Cazal, G.; Mbeguie-A-Mbeguie, D.; Ravion, S.; Pallet, D.; Abel, H. Changes in antioxidant activity during the ripening of jujube (*Ziziphus mauritiana* Lamk). *Food Chem.* **2014**, *150*, 448–456. [CrossRef] [PubMed]
25. Imran, M.; Khan, H.; Shah, M.; Khan, R.; Khan, F. Chemical composition and antioxidant activity of certain Morus species. *J. Zhejiang Univ. Sci. B* **2010**, *11*, 973–980. [CrossRef] [PubMed]
26. Zou, S.; Gao, P.; Jia, T.; Huang, H. Physicochemical Characteristics and Nutritional Composition during Fruit Ripening of *Akebia trifoliata* (Lardizabalaceae). *Horticulturae* **2022**, *8*, 326. [CrossRef]
27. Razzaq, K.; Khan, A.S.; Malik, A.U.; Shahid, M. Ripening period influences fruit softening and antioxidative system of 'Samar Bahisht Chaunsa' mango. *Sci. Hortic.* **2013**, *160*, 108–114. [CrossRef]
28. Zhao, Y.T.; Zhu, X.; Hou, Y.Y.; Pan, Y.F.; Shi, L.; Li, X.H. Effects of harvest maturity stage on postharvest quality of winter jujube (*Zizyphus jujuba* Mill. cv. Dongzao) fruit during cold storage. *Sci. Hortic.* **2021**, *277*, 109778. [CrossRef]
29. Saquet, A.A. Storage of pears. *Sci. Hortic.* **2019**, *246*, 1009–1016. [CrossRef]
30. Mahmood, T.; Anwar, F.; Abbas, M.; Boyce, M.C.; Saari, N. Compositional Variation in Sugars and Organic Acids at Different Maturity Stages in Selected Small Fruits from Pakistan. *Int. J. Mol. Sci.* **2012**, *13*, 1380–1392. [CrossRef]
31. Riaz, M.; Zamir, T.; Rashid, N.; Jamil, N.; Rizwan, S.; Masood, Z.; Mushtaq, A.; Tareen, H.; Khan, M.; Ali, M. Comparative study of nutritional quality of orange (*Citrus sinensis*) at different maturity stages in relation to significance for human health. *Am. Eur. J. Toxicol. Sci.* **2015**, *7*, 209–213.
32. Sanchez, E.M.; Calin-Sanchez, A.; Carbonell-Barrachina, A.A.; Melgarejo, P.; Hernandez, F.; Martinez-Nicolas, J.J. Physicochemical characterisation of eight Spanish mulberry clones: Processing and fresh market aptitudes. *Int. J. Food Sci. Technol.* **2014**, *49*, 477–483. [CrossRef]
33. Chen, M.Y.; Zhang, P.; Zhao, T.T.; Han, F.; Liu, X.L.; Zhong, C.H. Relationship between harvest indices and fruit quality traits in Actinidia chinensis 'Jintao'. *Plant Sci. J.* **2019**, *37*, 621–627.
34. Nardozza, S.; Gamble, J.; Axten, L.G.; Wohlers, M.W.; Clearwater, M.J.; Feng, J.Q.; Harker, F.R. Dry matter content and fruit size affect flavour and texture of novel Actinidia deliciosa. *J. Sci. Food Agric.* **2011**, *91*, 742–748. [CrossRef]
35. Khodabakhshian, R.; Emadi, B.; Khojastehpour, M.; Golzarian, M.R. Determining quality and maturity of pomegranates using multispectral imaging. *J. Saudi Soc. For. Agric. Sci.* **2015**, *16*, 322–331. [CrossRef]
36. Rolland, F.; Baena-Gonzalez, E.; Sheen, J. Sugar sensing and signaling in plants: Conserved and novel mechanisms. *Annu. Rev. Plant Biol.* **2006**, *57*, 675–709. [CrossRef]
37. Araya, T.; Noguchi, K.; Terashima, I. Effects of carbohydrate accumulation on photosynthesis differ between sink and source leaves of *Phaseolus vulgaris* L. *Plant Cell Physiol.* **2006**, *47*, 644–652. [CrossRef]
38. Sturm, K.; Koron, D.; Stampar, F. The composition of fruit of different strawberry varieties depending on maturity stage. *Food Chem.* **2003**, *83*, 417–422. [CrossRef]
39. Phillips, K.M.; McGinty, R.C.; Couture, G.; Pehrsson, P.R.; McKillop, K.; Fukagawa, N.K. Dietary fiber, starch, and sugars in bananas at different stages of ripeness in the retail market. *PLoS ONE* **2021**, *16*, e0253366. [CrossRef]
40. Lee, S.K.; Kader, A.A. Preharvest and postharvest factors influencing vitamin C content of horticultural crops. *Postharvest Biol. Technol.* **2000**, *20*, 207–220. [CrossRef]
41. Zhang, L.H.; Li, S.F.; Dong, Y.; Zhi, H.H.; Zong, W. Tea polyphenols incorporated into alginate-based edible coating for quality maintenance of Chinese winter jujube under ambient temperature. *LWT Food Sci. Technol.* **2016**, *70*, 155–161. [CrossRef]
42. Mditshwa, A.; Fawole, O.A.; Vries, F.; Kobus, V.D.M.; Crouch, E.; Opara, U.L. Impact of dynamic controlled atmospheres on reactive oxygen species, antioxidant capacity and phytochemical properties of apple peel (cv. Granny Smith). *Sci. Hortic.* **2017**, *216*, 169–176. [CrossRef]
43. Jin, P.; Wu, X.; Xu, F.; Wang, X.L.; Wang, J.; Zheng, Y.H. Enhancing Antioxidant Capacity and Reducing Decay of Chinese Bayberries by Essential Oils. *J. Agric. Food Chem.* **2012**, *60*, 3769–3775. [CrossRef] [PubMed]
44. Kirigia, D.; Winkelmann, T.; Kasili, R.; Mibus, H. Nutritional composition in African nightshade (*Solanum scabrum*) influenced by harvesting methods, age and storage conditions. *Postharvest Biol. Technol.* **2019**, *153*, 142–151. [CrossRef]
45. Zhang, Q.; Wang, L.L.; Liu, Z.G.; Zhao, Z.H.; Zhao, J.; Wang, Z.T.; Zhou, G.F.; Liu, P.; Liu, M.J. Transcriptome and metabolome profiling unveil the mechanisms of *Ziziphus jujuba* Mill. peel coloration. *Food Chem.* **2020**, *312*, 125903. [CrossRef] [PubMed]
46. Zoratti, L.; Karppinen, K.; Escobar, A.L.; Haggman, H.; Jaakola, L. Light-controlled flavonoid biosynthesis in fruits. *Front. Plant Sci.* **2014**, *5*, 534. [CrossRef]
47. Shwartz, E.; Glazer, I.; Bar-Ya'akov, I.; Matityahu, I.; Bar-Ilan, I.; Holland, D.; Amir, R. Changes in chemical constituents during the maturation and ripening of two commercially important pomegranate accessions. *Food Chem.* **2009**, *115*, 965–973. [CrossRef]
48. Fawole, O.A.; Opara, U.L. Changes in physical properties, chemical and elemental composition and antioxidant capacity of pomegranate (cv. Ruby) fruit at five maturity stages. *Sci. Hortic.* **2013**, *150*, 37–46. [CrossRef]
49. Koopman, W.J.; Nijtmans, L.G.; Dieteren, C.E.; Roestenberg, P.; Valsecchi, F.; Smeitink, J.A.; Willems, P.H. Mammalian mitochondrial complex I: Biogenesis, regulation, and reactive oxygen species generation. *Antioxid. Redox Signal.* **2010**, *12*, 1431–1470. [CrossRef]
50. Laxman, S.; Sutter, B.M.; Wu, X.; Kumar, S.; Guo, X.; Trudgian, D.C.; Mirzaei, H.; Tu, B.P. Sulfur amino acids regulate translational capacity and metabolic homeostasis through modulation of tRNA thiolation. *Cell* **2013**, *154*, 416–429. [CrossRef]

51. Buck, M.D.; Sowell, R.T.; Kaech, S.M.; Pearce, E.L. Metabolic instruction of immunity. *Cell* **2017**, *169*, 570–586. [CrossRef]
52. Zhang, X.; Wang, X.; Yuan, Z.; Radford, S.J.; Liu, C.; Libutti, S.K.; Zheng, X.F.S. Amino acids-Rab1A-mTORC1 signaling controls whole-body glucose homeostasis. *Cell Rep.* **2021**, *34*, 108830. [CrossRef]
53. Kelly, B.; Pearce, E.L. Amino assets: How amino acids support immunity. *Cell Metab.* **2020**, *32*, 154–175. [CrossRef]
54. Zhu, S.R.; Yang, G.; Yang, D.; Li, L.b.; Mo, R.J.; Li, X.L.; Zhang, Y.F. Comparative analysis of organic nutrients between *Holboellia latifolia* and nine common fruits. *J. Kunming Univ.* **2022**, *44*, 70–74.
55. Yang, Y.X. *China Food Composition Tables*, 6th ed.; Peking University Medical Press: Beijing, China, 2018.

horticulturae

Review

Prospects of Hogweed (*Heracleum sphondylium* L.) as a New Horticultural Crop for Food and Non-Food Uses: A Review

Eleonora Matarrese [1] and Massimiliano Renna [2,*]

1 Pikniq, Azienda Agricola & Home Food, Via Per Comnago 79, 28833 Brovello-Carpugnino, Italy
2 Department of Soil, Plant and Food Sciences, University of Bari Aldo Moro, Via Amendola 165/A, 70126 Bari, Italy
* Correspondence: massimiliano.renna@uniba.it

Abstract: *Heracleum sphondylium* L., commonly known as hogweed, common hogweed, or cow parsnip, is an herbaceous plant of the Apiaceae family native to Europe and Asia. This wild edible plant is common in grasslands, herb-rich meadows, hedges, wooded areas, roadsides, and railway embankments and in both waste and cultivated grounds. This review presents both the characteristics and ethnobotany, as well as the findings, technical advances, and potential of hogweed research with the goal of improving and disseminating knowledge regarding the value and potential of this wild edible plant. Current knowledge suggests that *H. sphondylium* L. shows good potential as a new cash crop, being an interesting food ingredient and also a source of compounds with biological activities. Therefore, hogweed may be proposed as a new horticultural crop, although several aspects of cultivation must be examined before full domestication.

Keywords: Apiaceae; biodiversity enhancement; biology; domestication; ethnobotany; food exploitation; marginal areas; phytochemistry

Citation: Matarrese, E.; Renna, M. Prospects of Hogweed (*Heracleum sphondylium* L.) as a New Horticultural Crop for Food and Non-Food Uses: A Review. *Horticulturae* **2023**, *9*, 246. https://doi.org/10.3390/horticulturae9020246

Academic Editors: Roberta Bulgari, Ada Baldi, Anna Lenzi and Antonios Chrysargyris

Received: 9 January 2023
Revised: 5 February 2023
Accepted: 9 February 2023
Published: 11 February 2023

1. Introduction

In the past, the collecting of wild edible plants (WEP) was the only option for survival during famines and chronic poverty [1–4], and for this reason, they are also known as "phyto-alymurgic plants" [5]. Today, WEP can be considered a great historical and cultural heritage that can restore a link to agrobiodiversity [6] and old gastronomic traditions [7] and improve diets [8]. Several studies have been carried out with the aim of cultivating certain WEP such as *Asparagus acutifolius* L. [9], *Borago officinalis* L., *Taraxacum officinalis* L. [10], *Muscari comosum* (L.) Mill. [11], and *Brassica fruticulosa* Cyr. [12]. Apart from these examples, an estimated 30,000 plant species are considered edible; however, nowadays, very few of them are grown as crops or are cultivated on a commercially significant scale.

The genus *Heracleum* is one of the largest of the Apiaceae family, including about 125 species. Among them, *Heracleum sphondylium* L. (Figure 1) occurs in most of Europe and parts of Asia and North Africa [13]. This species, commonly known as hogweed, common hogweed, or cow parsnip, is a perennial or biennial herbaceous plant of the Apiaceae family native to Europe and Asia. It is also called "eltrot"; however, this is not a specific common name for this species [14]. The American species *H. maximum* (also called cow parsnip) is sometimes included as a subspecies of *H. sphondylium* L.

The morphological similarity of the species within the genus and the difficulty of botanical identification have led to several synonyms and naming issues. For example, the classification of the species now widely known as *H. maxima* has been inconsistent. In literature, the scientific names *H. lanatum*, *H. maximum*, and others are used interchangeably. Before the 2000s, the previous name was more popular; today, the middle name is more popular.

Figure 1. Plants of hogweed (*Heracleum sphondylium* L.). Picture by Eleonora Matarrese.

In some parts of the world, *H. sphondylium* is used as an ingredient for several traditional recipes. For example, borscht derives from an ancient soup originally cooked with stems, leaves, and umbels of common hogweed, which resulted in its Slavic name [15]. The use of this species for a liqueur preparation in France and as food or a food additive in some Asian countries has also been reported [16]. Furthermore, this species is used in traditional medicine as an aphrodisiac, vasodilator, tonic, antihypertensive, and sedative and to treat dysentery, dyspepsia, digestive, and gynaecological problems [17].

Given the increasing scientific interest in *H. sphondylium* (Figure 2), in this review, its ethnobotany, characteristics, and potential will be presented with the aim of spreading knowledge and prospects for its cultivation as a new cash crop.

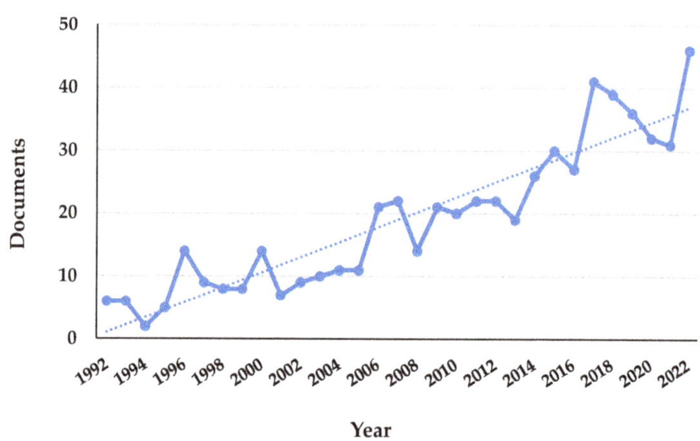

Figure 2. Increasing number of articles on *H. sphondylium* published from 1992 to 2022 (data retrieved from Scopus® database). The dashed line indicates the growing trend of the published articles.

2. Methodology

We could not identify any previous reviews focusing on *H. sphondylium*. Therefore, our review provides an interdisciplinary and international overview of this species. In the first step, we selected our research questions, the bibliographic article databases and websites, as well as the appropriate search terms. Then, we used practical review criteria for the inclusion or exclusion of the relevant literature. In the third step, we developed and applied methodological review criteria. Finally, we grouped data and information for elaboration in the different sections of the review.

2.1. Step 1: Selecting Research Questions, Databases, Websites, and Appropriate Search Terms

Because we could not identify any comprehensive articles on *Heracleum sphondylium*, our research questions for the review were rather broad: "How is *H. sphondylium* used?" or "What are the characteristics of *H. sphondylium*?". To search the literature, we chose the search terms "Heracleum AND sphondylium" as key terms for searching within the article title, abstract, and keywords by using the Scopus database. We also used Google Scholar to identify unpublished studies, conference proceedings, and similar publications following the recommendation of Tranfield et al. [18] that searches should not be restricted to bibliographic databases. Using the mentioned search terms, we searched the full text of all documents.

2.2. Step 2: Applying Practical Screening Criteria

We included journal papers, books and book chapters, research reports, and conference papers by accepting empirical publications as well as conceptual/theoretical publications. Quality criteria, such as journal rankings, were not used for exclusion purposes because this review aims to give a comprehensive overview of *Heracleum sphondylium*.

2.3. Step 3: Applying Methodological Screening Criteria

Within the third step, a review protocol for the content analysis of the publications was determined. The review protocol encompassed three sections. The first section contained the bibliographic data of each publication, such as author(s), year and title of the publication, authors' affiliations, type of publication, and, if it is a journal, the journal's name. The second section described the methodology of the publication: theoretical/conceptual, empirical, or practical-solution oriented. The third section focused on the content of the publication. For example, if the subject regarded "chemical analysis", we graded the document as a key source of chemical characterisation information. If the authors described the "uses", we graded the document as a key source of ethnobotanical knowledge. Moreover, we separately recorded which terms were included (i.e., aromatic compounds, furocoumarins, etc.).

In Step 4 we grouped data and information to elaborate on further in the next sections, also including authors' opinions, especially opinions regarding the prospects of this species.

3. Species Description

3.1. Distribution and Habitat

H. sphondylium is a plant with distribution in Europe south of 61° latitude, including Great Britain, up to north-western Africa, and both western and northern Asia. The species has also been introduced to suitable habitats elsewhere, such as in North America [19]. Throughout Europe, the distribution is broadly within mean annual temperatures of 5–15 °C and mean annual precipitation of 700–5000 mm. The Mediterranean region represents the southern limit to the distribution of this species, probably because the minimum winter temperatures are too high for the after-ripening requirements of the seeds (Section 7), and the extended drought periods affect seedling survival. In northern Europe, the distribution of this species may be limited by low temperatures, soil fertility, and pollination requirements. Its most common habitat is that of meadows and woods, especially in mountainous areas up to 2500 m, and is an indicator of soils rich in nitrogen [20]. Table 1 reports the classification of common hogweed habitats according to the CORINE biotopes classification [21].

Table 1. Habitat classifications of *H. sphondylium* according to the data specifications retrieved from the CORINE biotopes manual [21].

Code	Legend	Habitat Description
16	Coastal sand dunes and sand beaches	Sand-covered shorelines in general, but in particular, onshore areas of sand created by the action of wind and often colonized and stabilized by communities of coarse maritime grasses.
31	Heath and scrub	Temperate shrubby areas: Atlantic and alpine heaths, subalpine bush and tall herb communities, deciduous forest recolonisation, and hedgerows.
34	Dry calcareous grasslands and steppes	Dry thermophilous grasslands of the montane zone on mostly calcareous soil surfaces.
36	Alpine and subalpine grasslands	Grasslands of the alpine and subalpine levels of the Alps, Pyrenees, Cantabrian range, Jura, Central Massif, and northern Apennines, with very fragmentary outposts in the great Hercynian ranges of middle Europe, and in the Caledonian system of Britain; grasslands of the oro- and cryoro-Mediterranean levels or of the alti-Mediterranean level of the Iberian mountains and of the Apennines.
38	Mesophile grasslands	Lowland and montane mesophile pastures and hay meadows.
41	Broad-leaved deciduous forests	Forests and woodlands of native deciduous trees, other than floodplain or mire woods; forests dominated by broad-leaved deciduous trees, but comprising broad-leaved evergreen trees, are included.
42	Coniferous woodland	Forests and woodlands of native coniferous trees other than floodplain and mire woods; formations dominated by coniferous trees, but comprising broad-leaved evergreen trees, are included.
45	Broad-leaved evergreen woodland	Mediterranean forests dominated by broad-leaved evergreen trees. Laurel forests of the Atlantic islands. Holly woods.

The broad habitat range of *H. sphondylium* is largely a result of land management activities. The most "natural" habitats include forest or woodland clearings, coastal cliffs and dunes, riverbanks, and tall herb montane grasslands. Common hogweed occurs at a low density in most deciduous and some coniferous forests. In coastal grassland, *H. sphondylium* grows near the coastal fringe. It is absent from most maritime grasslands and saltmarshes, except on the non-waterlogged extreme landward side. Common hogweed is widespread on *Ammophila arenaria* dunes, especially in the north and in east Britain. It is characteristically absent from unimproved rank chalk grasslands and from rank grasslands on soils with a pH of less than approximately 4-3. Typical habitats also include hedgerows, road verges, wastelands, meadows, and abandoned pastures [22].

3.2. Morphology and Biology

The common hogweed plant develops from rhizomatous roots with a furrowed stem and bristly hair, growing up to 2 m high (Figure 3A,B).

The leaves are lobed and pinnate with small, serrated segments (Figure 3C). Hogweed has five-petalled pinkish or white flowers arranged in umbels, usually less than 30 cm in diameter, with 15 to 30 rays. The peripheral flowers have a radial symmetry (zygomorphic). The terminal umbels are flat-topped, and the outermost petals are enlarged. Flowering occurs between June and October. The flowering is phased within the umbel, with the outer row of larger flowers on each umbellet opening first. Self-pollination may occur, but the andromonoecious flowers are protandrous, tending to limit this. Anthesis precedes stigma receptivity within any flower by approximately 8–10 days. Geitonogamy is permitted, however, as filament length is relatively long [22]. The earliest flowers to open (i.e., those on the primary umbel) are the most likely to set seed. Most plants show facultative early abortion of carpels. This may result from the over-production of flowers that, due to limited plant resources, never set seed. The flowers are pollinated by insects, such as beetles, wasps, and especially flies. The small fruits are schizocarps, flattened and winged, elliptical to rounded and glabrous, and up to 1 cm long. The seed dispersal is by wind (anemochory) [20].

Figure 3. Plant details of *H. sphondylium*: stem (**A**,**B**); leaves (**C**); and shoots (**D**). Pictures by Eleonora Matarrese.

The typical number of annual seeds produced per plant is highly variable, on average about 850. Germination is epigeal, but the seeds require about eight weeks' moist after-ripening at <2 °C before they are ready for germination. The cold germination requirement may limit the spread of *H. sphondylium* in Southern Europe. The seedlings have panduriform cotyledons. Subsequently, the seedlings develop a rosette of mature leaves and a substantial taproot weighing 5–30 g by the beginning of the second year [22].

Nine subspecies may be recognized: ssp. *alpinum* (L.) Bonnier and Layens, *montanum* (Schleicher ex Gaudin) Briq., *orsinii* (Guss.) H. Neumayer, *pyrenaicum* (Lam.) Bonnier and Layens, *sybiricum* (L.) Simonkai, *sphondylium*, *ternatum* (Velen.) Brummitt, *transsilvanicum* (Schur) Brummitt, and *verticillatum* (Pancic') Brummitt. Two subspecies are native to Britain: *H. sphondylium* ssp. *sphondylium* and *H. sphondylium* ssp. *sybiricum*. Other subspecies are found chiefly in the Balkans and in the mountains of Europe [22].

Do not confuse the smaller *H. sphondylium* with the dangerous *H. mantegazzianum* (giant hogweed) or *H. sosnowskyi* (Sosnowsky's hogweed). Giant hogweed typically grows up to 5 m. A mature plant has huge leaves, between 1–1.5 m wide, and a stout, bright green stem with extensive dark reddish-purple splotches and prominent coarse white hairs, especially at the base of the leaf stalk. *H. sosnowskyi* is also smaller than giant hogweed (growing up to 3 m) and is more commonly found in northern areas of Europe; it is more resilient to harsh conditions than the other two species. The clear watery sap of giant hogweed contains toxins that can cause severe dermatitis (inflammation of the skin). Ultraviolet radiation activates compounds in the sap resulting in severe burns when exposed to the sun. Symptoms occur within 48 h and consist of painful blisters.

H. sphondylium shows a high tolerance to wind exposure and atmospheric salinity. However, plants near the coastal fringe can form semi-succulent leaves due to salt spray, while exposure to grazing and wind reduces the plant to hard stunted rosettes with short leaves [22]. Plants can associate with vesicular-arbuscular mycorrhizas, which have a beneficial role in nutrient uptake, although these associations are facultative [23].

Light requirements are low, considering that *H. sphondylium* is a common deciduous woodland species, up to light levels of 5% daylight on the forest floor [22].

4. Ethnobotanical Knowledge

The specific term *sphondylium* means "vertebrate" and refers to the shape of its segmented stem. It was described by Linnaeus in 1753 [3]. Heracleum, on the other hand, is the name of its genus; it derives from ancient Greek ʽΗράκλειος (Hērákleios), "of Hera-cles", with reference to the mythological hero.

The characteristic "farmyard" smell or the observation that pigs can eat hogweed's foliage and roots are perhaps the origins of its common name in the English language [24]. *Hog*, "pig", plus *weed*, entered the language in 1707; used in various different places for plants eaten by pigs or thought to be suitable for them only.

The small fly *Euleia heraclei*, known as the "celery fly" or the hogweed "painted-winged fly", is a species of tephra or fruit fly in the genus *Euleia* of the Tephritidae family, which is found as the binomial specification suggests on hogweed [25].

Ethnobotanical uses of *H. sphondylium* L. are summarized in Table 2.

Table 2. Literature regarding the ethnobotanical uses of *H. sphondylium*.

Product Type	Uses	References
Herbal tea of aerial parts	Aphrodisiac and against hypertension	Senejoux et al. [13]
Root decoction	Against dysentery	Işcan et al. [16]
Distillate of aerial parts	Liqueur	Heather [26]
Whole stems, leaves, and umbels	Borscht (ancient soup)	AA.VV [19]
Extract of aerial parts	Against gynaecological and fertility problems	AA. VV. [19]
	Sedatives of the nervous system	Colombo and Luciano [27]
	Antidepressant	Colombo and Luciano [27]
Root infusion	Against impotence and frigidity	Colombo and Luciano [27]

In Romania and Morocco, the herbal tea of the aerial parts of hogweed is reputed to be an aphrodisiac and able to treat hypertension [13]. *H. sphondylium* is known as *tavşancılotu* and used against dysentery in Turkey [16]. In the 18th century, the inhabitants of the Kamchatka Peninsula distilled a liquor called *raka* from a "sweet herb" believed to be *H. sphondylium* [26]; raka was then flavoured with blue berried honeysuckle (*Lonicera caerulea*).

Borscht is a sour soup common in Eastern Europe and Northern Asia. In English, the word "borscht" is most often associated with the soup's variant of Ukrainian origin, made with red beetroots as one of the main ingredients, which gives the dish its distinctive red colour [15]. This dish derives from an ancient soup that was originally prepared by using hogweed, which lent the dish its Slavic name. Growing commonly in damp meadows throughout the north temperate zone, hogweed was used not only as fodder (as its English names suggest) but also for human consumption—from Eastern Europe to Siberia to northwestern North America. The name borscht ultimately derives from the word борщ (*borshch*), which is common to East Slavic languages, such as Ukrainian. Together with cognates in other Slavic languages, it comes from Proto-Slavic **bŭrščĭ*, "hogweed", and ultimately from Proto-Indo-European **bhr̥stis*, "point, stubble" [15]. Common hogweed was the soup's principal ingredient before it was replaced with other vegetables, notably beetroot in the Ukrainian version. The English spelling comes from Yiddish (borsht), as the dish was first popularized in North America by Yiddish-speaking Ashkenazi Jews from Eastern Europe. With time, borscht evolved into a diverse array of tart soups, among which the Ukrainian beet-based red version has become the most popular [15]. Its popularity has spread throughout Eastern Europe and—by way of migration away from the Russian Empire—to other continents. In North America, borscht is often linked to either the Jewish or Mennonites, the groups who first brought it there from Europe. Several ethnic groups claim borscht, in its various local guises, as their own national dish consumed as part of ritual meals within Eastern Orthodox, Greek Catholic, Roman Catholic, and Jewish religious traditions [15].

In Eastern European countries and, in particular, Romania, *H. sphondylium* is used as an aphrodisiac and to treat gynaecological, fertility, and impotence problems. It is also sometimes recommended for epilepsy [19]. In Piedmont, "potions" were commonly prescribed as sedatives for the nervous system. In cooking, spring sprouts were and are used raw in salads or cooked like asparagus. From its seeds, they prepare a liqueur with a pleasant taste [27].

The hogweed was a very common and famous herb in the Renaissance period for fighting depressive crises. The infusion of the root against impotence and frigidity has been known since ancient Egypt [27].

A decoction of roots of the similar species *H. asperum* or *H. leskovii* was prepared in Georgia by a local healer in Svaneti as a remedy to purify the body and also used to cure cancer. *H. wilhelmsii*, another similar species, was foraged, and its roots were used to treat stomach problems [28].

5. Food Uses

Hogweed must be harvested as soon as it blooms but can be dried for later use. It can be used as a fresh vegetable by harvesting its shoots (Figure 3D) as soon as they sprout and before flowering and used in soups or salads. It tastes like asparagus; the crust is somewhat acrid. The stems of the leaves are tied in bunches and dried in the sun until they turn yellow. From these, a sweet substance similar to sugar is formed and is considered a great delicacy. The root can also be eaten cooked; it is usually boiled. This plant can be prepared as an infusion, decoction, or tincture. Hogweed's young shoots are considered excellent food by many foragers. Hogweed seeds are used as a spice: their flavour is reminiscent of cardamom. If they are harvested unripe, their flavour is reminiscent of mandarins and clementines, with a strong citric scent. The seeds can be macerated in alcohol to prepare a pleasant liqueur. A tincture made from the aerial parts of the plant has also been used to relieve general weakness [20].

In Georgia, *Heracleum* spp. is harvested and eaten raw, pickled, fermented, and eaten fresh, including the peeled root. The young, fresh stems, called *k'ap'i* (singular) and *k'ap'ebi* (plural), are eaten before flowering. The young shoots, just sprouted in spring, are used as fresh food in various alpine regions of western and eastern Georgia. The same is eaten in sour milk (during summer in a recipe called *sats'eba*). Its leaves have also been put in sour milk and eaten as raw soup. The stems of *H. asperum*, a similar species, are peeled, and its fresh inner part is eaten, which is said to have a sweet taste. Tradition has it that the *shup'q'a*, its internal part, had to be eaten within 24 h because it withers the next day [28].

To date, following our gastronomic experiments aimed at the culinary enhancement of *H. sphondylium*, this species can be proposed as an ingredient in sweet and savoury dishes: hogweed bread (Figure 4); hogweed *gelo* with percoca peach and wild strawberries in syrup (Figure 5); hogweed and burdock 'fake artichokes' (Figure 6); and "Amaretti" with *H. sphondylium* (Supplementary Materials). Furthermore, thanks to broader ethnobotanical knowledge, we hypothesized several potential uses for hogweed likened to other wild species, especially of the Apiaceae family.

Figure 4. Hogweed bread. Picture and recipe by Eleonora Matarrese. The full recipe is available in the Supplementary Materials.

For example, like *Angelica archangelica*, hogweed can be used to flavour liqueurs and spirits (such as Chartreuse, Bénédictine, Vermouth, and Dubonnet), flavour fish, and in the preparation of absinthe, brandy, and bitters, as well as other culinary uses, such as jams, compotes, jellies, omelettes, and fillings. Its stems, especially if large enough, can be eaten as they are or added to salads; without leaves, they can be crystallized in sugar syrup and used as a decoration for cakes or to make candies.

Hogweed's fruits can be used in an infusion to stimulate appetite and facilitate digestion in the same way as the wild carrot (*Daucus carota* L.). While, like *Torilis japonica*, its roots can be peeled and cooked, either steamed or boiled, and eaten as a vegetable; they can also be used in purées, stuffed, dried, and powdered.

Leaves and roots can be used as a seasoning and to flavour some kinds of cheese, as with *Peucedanum ostruthium*. Its root cooked in wine is still used today in Switzerland, while the roots and rhizome can be used to prepare aperitifs and liqueurs for digestive purposes. Like *Laserpitium latifolium*, hogweed can be used together with cumin to season preserved artichokes, as the ancient Romans did.

Hogweed's fresh leaves can replace both parsley and celery, as happens with *Ligusticum mutellina*, while dried leaves can be used as tea. Its seeds, with an intensely aromatic flavour, can be mixed with flour to prepare sweets and bread and in the preparation of some kinds of cheese, as with *Carum carvi*. They give a particular touch to meat dishes (especially pork), and they can also be used to prepare the well-known *kúmmel* liqueur, aromatic and digestive. Dried leaves of hogweed can also be used as a new spice colourant in culinary preparations, as with *Crithmum maritimum* L. [29,30].

Figure 5. Hogweed *gelo* with *percoca* peach and wild strawberries in syrup. Picture and recipe by Eleonora Matarrese. The full recipe is available in the Supplementary Materials.

Figure 6. Hogweed and burdock 'fake artichokes'. Picture and recipe by Eleonora Matarrese. The full recipe is available in the Supplementary Materials.

Like *Meum athamanticum*, the root of hogweed can be used as an appetite stimulant and digestive and to reduce menstrual pain. Roots can also be added as an ingredient to soups and stews, while leaves can be used to flavour several dishes. Roots can also be used raw, cut into julienne strips with extra virgin olive oil, salt, and apple cider vinegar (or lemon juice), or sautéed in a pan with garlic or chives, as well as in soups, risottos, omelettes, casseroles, quiches, and as a dip and in fillings, as with *Bunium bulbocastanum*.

Hogweed's fruits can be combined with porchetta and fatty meats, while its young shoots and leaves are indicated to flavour salads, fish, sauces, and aromatic vinegars, as made with *Foeniculum vulgare*.

Finally, like chervil (*Anthriscus cerefolium*), hogweed can be used as a flavouring in vinegar since it gives a very strong citrus scent. Thus, hogweed can also be used to make syrup to prepare refreshing drinks with a citrus flavour.

6. Phytochemistry and Biological Activity

Phytochemical investigations on *H. sphondylium* showed the presence, in seeds and roots, of furocoumarins (bergapten, isopimpinellin, and heraclenin) [31–33] and essential oils [34] (Figure 7). The essential oils consist mainly of monoterpenes, sesquiterpenes, and phenyl-propanoid compounds [22].

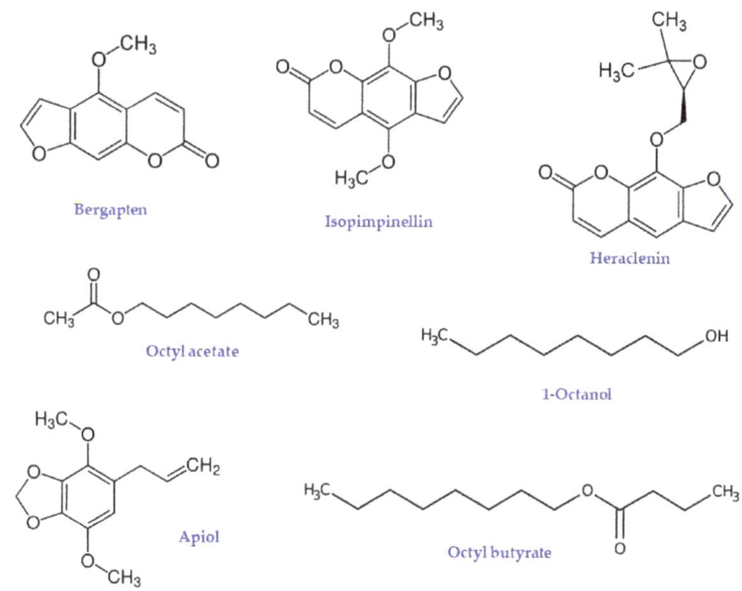

Figure 7. Chemical structure of some furocoumarins (Bergapten, Isopimpinellin, and Heraclenin) and volatile compounds (Octyl acetate, 1-Octanol, Apiol, and Octyl butyrate) found in *H. sphondylium*.

Furocoumarins are a class of chemical compounds with phototoxic properties found naturally in many plant species, including some commonly consumed by humans (i.e., grapefruit, carrot, parsnip, turnip, fig, lemon, lime, orange, celeriac, celery, parsley, and dill) [35]. They are synthesized by plants through the fusion of coumarin to a furan ring, generating linear or angular isomers depending on the position of the furan ring. Furocoumarins are produced by plants in response to stress and to defend against predators, such as fungi, bacteria, and insects; furocoumarins react with the DNA of these predators and disrupt replication when exposed to UV light [35]. Furocoumarins have been the focus of much research attention because of their photoactivity. Contact with furocoumarins combined with UV exposure can lead to the development of blistered and burned skin, a reaction known as phytophotodermatitis. However, both the ingestion of and dermal contact with furocoumarin-containing plants enables the absorption of furocoumarins into the bloodstream [35]. The concentration of furanocoumarins in *H. sphondylium* is much lower than that of the Caucasian hogweeds (*H. mantegazzianum*—giant hogweed and *H. sosnowskyi*—Sosnowsky's hogweed). However, there is evidence that the sap from common hogweed can also produce phytophotodermatitis when contaminated skin is exposed to sunlight [36].

Some biological activities of *H. sphondylium* L. are reported in Table 3.

Table 3. Literature regarding the biological activity of *H. sphondylium*.

Biological Activity	Product Type	References
Antibacterial	Essential oils	Iişcan et al. [16]
Antibacterial	Plant extract	Bahadori et al. [37]
Antifungal	Essential oils	Iscan et al. [38]
Antifungal	Plant extract	Uysal et al. [17] Bahadori et al. [37] Fierascu et al. [39]
Antioxidant	Plant extract	Benedec et al. [40] Uysal et al. [17] Fierascu et al. [39]
Inactivation of enzymes involved in Alzheimer's disease	Plant extract	Uysal et al. [17]
Bio-herbicide	Plant extract	Fierascu et al. [39]
Vasorelaxant	Plant extract	Senejoux et al. [13]
Cytotoxic against tumour cells	Essential oils	Maggi et al. [34]

The essential oil of *H. sphondylium* seeds, rich in 1-octanol and octyl butyrate, showed significant antimicrobial activity [16], while octyl acetate and octyl butyrate were the most abundant aliphatic esters identified in the essential oil of the fruits [34].

Senejoux et al. [13] investigated the vasorelaxant effects of a dichloromethane extract of *H. sphondylium* and the mechanisms involved. The authors showed that this extract exhibited vasorelaxant properties through endothelium-independent mechanisms involving the inhibition of Ca^{2+} mobilisation and changes in K^+ channels conductivity.

In a study aimed to evaluate the phenolic composition in different parts of the *H. sphondylium* plant, Benedec et al. [40] found high amounts of rutin in flowers (984 mg/100 g) and in leaves (477 mg/100 g). Regarding other flavonoids, quercitrin was found in leaves (15 mg/100 g), and quercetin in flowers (13 mg/100 g); ferulic acid (13.04 mg/100 g) and chlorogenic acid (4.32 mg/100 g) were found in roots. The same authors also found that flower and leaf extracts exhibited the highest antioxidant capacity, according to the phenolic content. In this respect, this plant shows immense potential to be domesticated and used industrially [17].

Uysal et al. [17] investigated the inhibitory action of different extracts of *H. sphondylium* on key enzymes involved in Alzheimer's disease, type 2 diabetes, and epidermal hyperpigmentation conditions. Additionally, these authors studied in silico molecular docking to provide additional insight into the interaction of the phenolic compounds identified in the different extracts with the analysed enzymes. The methanol extract, rich in chlorogenic acid, was revealed to be a good inhibitor of AChE. However, docking studies showed that rutin and quercetin possess high binding energy when docked to AChE. Thus, the authors concluded that the observed inhibitory action of the methanol extract on AChE might be due to the synergistic action of the phenolic compounds present [17]. AChE activity and oxidative stress play a major role in the onset and progression of Alzheimer's disease. In addition, the methanol extract showed strong antimutagenicity against 4-nitro-O-phenylenediamine, a powerful direct-acting mutagen, and 2-amino anthracene, a pro-mutagen [17]. The methanol extract also showed potent antioxidant properties on a battery of in vitro assays as well as moderate antifungal activity against *Candida albicans* and *C. parasilopsis* [17].

Antimicrobial activity was also described by other authors. Effectively, the ethanol and aqueous extracts of *H. sphondylium* showed antibacterial activity against *Staphylococcus aureus*, *Enterococcus faecalis*, *Escherichia coli*, *Pseudomonas aeruginosa*, *Listeria monocytogenes*, *Shigella*, *Streptococcus pyogenes*, and *Corynebacterium diphtheriae*, as well as antifungal activity against *Candida albicans* and *C. krusei* [37].

The essential oils of *H. sphondylium* showed good antifungal activity against *Candida glabrata*, highlighting an inhibitory effect higher than the antifungal agent ketoconazole [38].

Fierascu et al. [39] found that the hydroalcoholic extract of *H. sphondylium* showed good antioxidant potential and antifungal activity against *Aspergillus niger* and *Penicillium hirsutum*, suggesting its use as a natural antifungal agent for the treatment of fruits and vegetables in the postharvest period. The results of this study also suggested the use of the extract as a bio-herbicide [39].

In a study aimed to evaluate the effect of the essential oil of *H. sphondylium* against four human tumour cell lines, Maggi et al. [34] found that the essential oil showed moderate cytotoxic effects against a human malignant melanoma cell line and a human colon carcinoma cell line.

7. Domestication

Both ethnobotanical knowledge and food use suggest a good potential for the domestication of hogweed, although literature information relating to its cultivation is very scarce and fragmentary.

Seeds of *H. sphondylium* show an underdeveloped embryo (morphological dormancy—MD), as well as physiological inhibiting mechanisms of germination (physiological dormancy—PD), and thus, it can be described as a species with morphophysiological (MDP) dormancy. In particular, Baskin and Baskin [41] described common hogweed as a species with deep complex MPD, and its seeds would require only cold stratification for loss of PD and MD of the embryo. Effectively, Stokes [42] demonstrated that low temperatures were necessary to make food reserves in the endosperm of *H. sphondylium* seeds available to the growing embryo. The author found that seeds of *H. sphondylium* need a period of 9–12 weeks at 2–5 °C to germinate, while at 15 °C, seeds remain dormant. This is because low temperatures stimulate the rapid hydrolysis of endosperm proteins into soluble nitrogenous compounds and the formation of the amino acid glycine and arginine, which, together with soluble sugars, are beneficial for the growth and germination of embryos. On the other hand, at high temperatures (15 °C), soluble nitrogenous compounds are not available, and alanine, which does not stimulate embryo growth, results as the most abundant amino acid [42].

It is important to highlight that the application of gibberellic acid did not stimulate the germination of seeds of *H. sphondylium* [41]. Therefore, the domestication of hogweed must include a protocol for cold stratification of its seeds in order to overcome dormancy and optimize the percentage of germination. To this end, future research activities could carry out protocols already used for similar species starting from propagation. For example, the propagation protocol for the production of *H. maximum* provides a treatment of a 72 h water soak of fresh seeds, with the water changed daily. Then, seeds are placed into 100 days of cold, moist stratification, i.e., in fine mesh bags and buried in moist peat moss in a ventilated container under refrigeration at 1 to 3 °C [43]. This protocol describes soaking in water as necessary to leach out inhibitors on seed coats. It is well known that breaking MPD requires embryo growth and/or differentiation, and the seed must be imbibed for this to happen. It is not a case that MPD is frequent in parts of the world with moist seasonal climates and particularly common in plants (i.e., *Heracleum* genera) of woodlands or damp grasslands. In our opinion, for *H. sphondylium*, the seeds soaking in water could also be compared to other treatments, such as their priming into solutions containing glycine and arginine, with the aim to verify a possible faster overcome of dormancy. Furthermore, future research activities could be aimed at evaluating the effects of the photoperiod, light intensity, and different temperatures on hogweed seed germination, as for other species of the *Apiaceae* family [44,45].

H. sphondylium usually grows in fertile, clay-rich soil with a pH of 6–7, although it can also grow in soils with a pH of about 4.3, provided that they contain adequate amounts of nitrogen, phosphorous, and potassium [22]. Effectively, this species is typically absent on chalk heath and limestone heath soils and also from soils on dolerite and basalt since the availability of nitrogen and phosphorous in these soils is low [22]. Williams [46] first found that the nutrient status of the soil is the most important factor affecting the distribution and yield of this species; the author observed greater biomass levels of hogweed when the soil was enriched using a potassium-based fertilizer. In this regard, future research activities could be carried out to evaluate the effect of fertilisation techniques on the yield and quality of this species. Indeed, some aspects, such as the optimal amounts of fertilizers per area and the different forms of nitrogen, need to be studied with the aim of full domestication of this species. At the same time, we think that the possible implementation of sustainable management of fertilisation (i.e., organic fertilisation or living mulch) [47,48] would be desirable in the context of organic farming and, more generally, from an environmentally friendly point of view. On the other hand, since data about open-field cultivation of hogweed is not present in the literature, other growing aspects, such as plant density and irrigation, should be necessarily studied by experimental trials.

Apart from clay-rich soils, *H. sphondylium* also grows on the fertile brown loams of Scottish mountains and in continental Europe, in soils rich in nitrogen, as well as in soils of coastal habitats, with moderate salinity derived from sea spray and not by saltwater from inundation [22]. The different habitats in which hogweed grows spontaneously suggest the possibility of its cultivation in marginal areas, from the coast to mountainous areas. In this context, a way to domesticate hogweed without exploiting forest lands could be the forest farming of this species; that is, its cultivation in the forest understory of either established or developing forests [49]. To this end, it is important to cultivate plants that can grow under a forest canopy without negative effects due to the shade of the trees. In a study aiming to evaluate the effects of shade and planting methods on the growth of *Heracleum moellendorffii* [50], the authors found that shading significantly improved the height growth of the plants (10–20 cm increase) in both unfertilized and fertilized plots with highly moist soil conditions. Results also highlighted that shading improved aboveground production in unfertilized plots, in agreement with the characteristics typical of shade-tolerant species [50]. According to Sheppard [22], light requirements are low for *H. sphondylium,* although shaded plants have low seed production, producing tall elongate individuals with relatively few umbels. Therefore, future research activities could be aimed at evaluating the effects of shading and other treatments on the growth and yield of *H. sphondylium,* as for other species of the same botanical genera [50].

However, we cannot overlook the potential risks regarding the domestication of *H. sphondylium* in the open field due to possible hybridisation between common hogweed and Caucasian ones, especially *H. mantegazzianum* [51–53]. Giant hogweed is native to the western Caucasus region of Eurasia. It was introduced to Britain as an ornamental plant in the 19th century and has also spread to other areas in Western Europe, the United States, and Canada. Its close relatives, Sosnowsky's hogweed and Persian hogweed, have similarly spread to other parts of Europe. It is important to highlight that the sap of giant hogweed is phototoxic and causes phytophotodermatitis in humans, resulting in blisters and scars. These serious reactions are due to the furanocoumarin derivatives in the leaves, roots, stems, flowers, and seeds of the plant. Consequently, it is considered to be a noxious weed in many jurisdictions.

To this end, the domestication of common hogweed by using a controlled environment (such as greenhouses) could be an effective way to reduce the risk of hybridisation between common hogweed and Caucasian hogweeds. Moreover, the evaluation of soilless cultivation systems for the full domestication of hogweed could also be useful since several studies report the application of these systems to wild edible species. For example, the domestication of *Urospermum dalechampii* and *U. picroides* was tested for the first time, using a floating system to evaluate the yield and quality of these species as potential new vegetables for the ready-to-eat production chain [54]. Montesano et al. [55] evaluated the cultivation of potted *Crithmum maritimum* L., a wild edible halophyte of the Apiaceae family, applying a closed cycle ebb and flow hydroponic system and using a seaweed-based compost as a sustainable peat substitute for the formulation of media mixtures, even up to a complete peat replacement. The protocol for the production of potted *H. maximum* provides a growing medium composed of 50% milled sphagnum peat, perlite, and vermiculite with controlled release fertilizer for macronutrients (13N:13P2O5:13K2O) and fertilizer for micronutrients (12% S, 0.1% B, 0.5% Cu, 12% Fe, 2.5% Mn, 0.05% Mo, and 1% Zn) [43]. Soilless cultivation systems were primarily developed in response to the excessive spread of soil pathogens; however, they also allow for optimal control of plant growth, high productivity, and product quality, as well as very highly efficient water and fertilizer use. At the same time, consumers remain critical towards soilless-cultivated vegetables, mainly due to the perception of these techniques as unnatural, resulting from artificial growth and consequently characterized by low quality [56]. In our opinion, for *H. sphondylium*, like for other Apiaceae species, the experimental evaluation of alternative growing media to peat and perlite could be a way towards the domestication and, at the same time, increased sustainability of soilless systems in the perspective of a circular economy [56].

Some studies also proposed the application of soilless cultivation systems for the biofortification of wild edible species. For example, D'Imperio et al. [57] described the boron biofortification of *Portulaca oleracea* L. through soilless cultivation for a new tailored crop. An increase in the nutritional value of *P. oleracea* L., also through biofortification with zinc [58] and silicon [59], was carried out using a floating system. Puccinelli et al. [60] described the selenium biofortification of *Rumex acetosa* L., *Plantago coronopus* L., and *P. oleracea* L., grown as microgreens. Other authors investigated the effect of selenium on the growth, yield, and biofortification of *Eruca sativa* L. grown in a hydroponic system [61]. Biofortification is a process that can be applied to fresh, uncooked vegetables with the primary aim of improving their nutritional quality. Considering the increasing research activities that have been directed towards this technique in the last few years [62], the prospect of soilless cultivation for *H. sphondylium*, and also for biofortification objectives, could be a further option in the context of biodiversity enhancement.

The soilless cultivation of wild edible plants under an agrivoltaic greenhouse represents another important research topic. The process of co-developing photovoltaic (PV) electricity generation and crop cultivation on the same land is called "agrivoltaics", where the prefix "agri" refers to the science and technology of producing crops in agriculture, and "voltaic" refers to PV power generation. To this end, Buttaro et al. [63] evaluated the soilless production of the wild rocket (*Diplotaxis tenuifolia* L.) as affected by greenhouse coverage with PV modules. In this study, the authors suggested the possibility of combining soilless cultivation and solar energy production, highlighting the importance of choosing species that are not negatively affected by shading in terms of the yield and quality of the cultivated product. Effectively, the integration of traditional opaque PV modules in crop cultivation environments caused adverse impacts on crop growth due to the shadow effect, especially when high shading ratios have occurred. *H. sphondylium* can be considered a species with low light requirements [22]. Therefore, we hypothesize that future research activities can also be aimed at evaluating the adaptability of common hogweed to cultivation under agrivoltaic systems.

8. Prospect

Apart from the traditional and innovative food uses of *H. sphondylium*, several biological activities (Figure 8) are reported in the literature regarding their non-food uses potentially being exploitable by the pharmaceutical and agri-food industries.

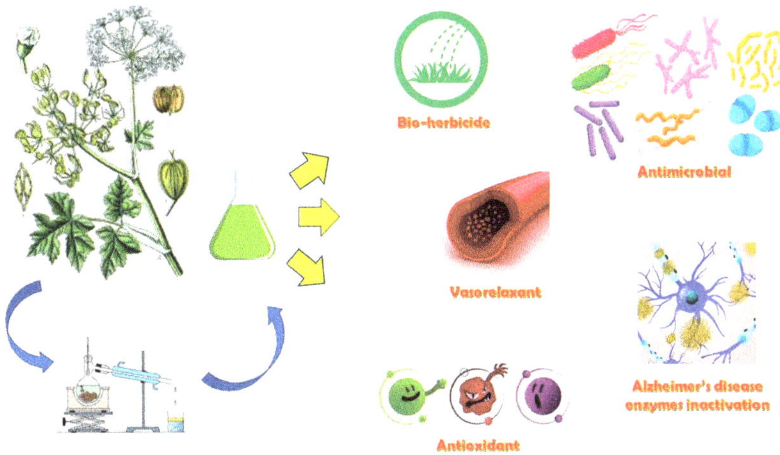

Figure 8. Graphical representation of the different biological activities attributable to *H. sphondylium*.

Therefore, in order to evaluate whether hogweed can be regarded as a new horticultural crop with a concrete chance to succeed, a SWOT analysis was performed. SWOT (Strengths, Weaknesses, Opportunities, and Threats) analysis comprises the analysis of the strengths and weaknesses of a project, product, place, or person and their relationship with the opportunities and threats of the surroundings. In short, it is a framework for identifying and analysing the internal and external factors that can have an impact on the viability of a project, product, place, or person. SWOT analysis is considered an important decision-making tool and is often used to systematically analyse the internal and external environments of projects, products, and organisations. Table 4 reports the SWOT analysis regarding the exploitation of hogweed as a new cash crop.

Table 4. SWOT analysis related to the exploitation of *H. sphondylium* as a new cash crop.

Strengths	Weaknesses
• Traditional vegetable • Good organoleptic traits • Richness in chemical compounds • Adaptability to different habitats	• Knowledge is limited to local areas, and a few researchers • Absence of market • Absence of specific cultivation protocols
Opportunities	**Threats**
• New products • Exploitation of marginal areas • Soilless cultivation • Consortia for R&D	• Resistance by consumers and farmers • Hybridisation

Regarding the strengths, it is important to first specify that hogweed can be considered an ancient and traditional wild vegetable used for making various dishes and processed products. This species of the Apiaceae family shows interesting organoleptic traits and a great diffusion in different habitats, from meadows and woods of mountainous areas to coastal areas, showing good adaptability to different pedoclimatic conditions. This strength makes hogweed a good candidate as a crop for the exploitation of marginal areas such as forest understory and coastal habitats. Of course, its richness in chemical compounds is not of secondary importance. Therefore, the development of a specific agri-food chain based on both fresh and processed hogweed products could favour the development of new markets able to meet the growing demand for functional products. However, it must be highlighted that: (i) there is currently no widespread market for hogweed, and (ii) hogweed knowledge is limited to local areas and a few researchers. These weaknesses, therefore, require specific activities to be carried out in order to disseminate knowledge, promote potential uses, and boost consumer demand. To this end, hogweed exploitation could require multi-disciplinary activities and integrated projects [64].

Thanks to its richness of useful chemical compounds, hogweed seems to be a very promising candidate for both the pharmaceutical and agri-food industries for the production of new functional products. At the same time, the presence of some specific compounds makes this species also interesting for the industrial production of natural herbicides and crop protection products for the post-harvest sector. At any rate, according to Petropoulos et al. [1], a multi-step approach could be needed for hypothesizing the full commercialisation of these new products. In this context, the evaluation of several hogweed populations from different geographical areas should be first carried out in order to select the best chemotypes for specific uses. At the same time, also needed could be an evaluation of the potential differences regarding the bioactive compounds content of plants under cultivation conditions with respect to wild plants, although currently, there are no tested protocols for its cultivation.

Some threats may arise due to the potential resistance of consumers and farmers regarding this species as a functional food and/or new cash crop. This may require a few preventive activities, including clinical trials for evaluating effects on health and a specific marketing project to achieve increased consumer acceptance. Therefore, the establishment of consortia between research institutes, business companies, and governmental organisations aiming to carry out these research and development activities could be a good opportunity. Other threats may arise due to possible hybridisation between common hogweed and the dangerous *H. mantegazzianum*. For this reason, the domestication of hogweed in controlled environments could be an effective way to reduce this risk. In this context, the application of soilless cultivation systems may be a further opportunity.

9. Conclusions

Current knowledge suggests that *H. sphondylium* shows good potential as a new horticultural crop, being a refined food and also an interesting source of health compounds, as well as for non-food products. Furthermore, hogweed could be an alternative for both horticultural and industrial crops in the presence of soils in marginal areas, from the coast to mountainous regions. This review also suggests that ethnobotany may offer a source of inspiration for agriculture, as hogweed and, more generally, all wild edible plants have the potential to lead food systems to be healthier, more sustainable, and resilient to climate change in the context of biodiversity enhancement. On the other hand, the interesting characteristics of hogweed and its several uses are known by researchers only through scientific literature and/or by a low percentage of people in niche areas. Therefore, a multi-disciplinary approach and integrated projects should be used for hypothesising the commercialisation of this potential new horticultural crop. Finally, several cultivation aspects will need to be examined before hypothesizing the full domestication of this wild edible plant.

Supplementary Materials: The following supporting information can be downloaded at: https://www.mdpi.com/article/10.3390/horticulturae9020246/s1, Hogweed recipes related to Figures 4–6, as well as to "Amaretti" with *H. sphondylium*.

Author Contributions: E.M. contributed mainly to Section 5. M.R. contributed mainly to Sections 7 and 8. Both authors contributed equally to the drafting of the other parts of the manuscript. All authors have read and agreed to the published version of the manuscript.

Funding: This study was carried out within the Agritech National Research Center (Spoke 7) and received funding from the European Union Next-GenerationEU (PIANO NAZIONALE DI RIPRESA E RESILIENZA (PNRR)—MISSIONE 4 COMPONENTE 2, INVESTIMENTO 1.4—D.D. 1032 17/06/2022, CN00000022). This manuscript reflects only the authors' views and opinions; neither the European Union nor the European Commission can be considered responsible for them. This research was also supported by the Regione Puglia Administration under Rural Development Program 2014–2020, Project 'Biodiversity of Apulian vegetable species (BiodiverSO Veg)', Measure 10, Sub measure 10.2, Operation 1 "Program for the conservation and the valorization of the genetic resources in agriculture" (DDS n. 04250182807, CUP: B97H22003760009)—n. 3.

Data Availability Statement: Not applicable.

Acknowledgments: Authors are grateful to Patrizia Renna Fray for manuscript editing.

Conflicts of Interest: The authors declare no conflict of interest.

References

1. Petropoulos, S.A.; Karkanis, A.; Martins, N.; Ferreira, I.C.F.R. Edible halophytes of the Mediterranean basin: Potential candidates for novel food products. *Trends Food Sci. Technol.* **2018**, *74*, 69–84. [CrossRef]
2. Licata, M.; Tuttolomondo, T.; Leto, C.; Virga, G.; Bonsangue, G.; Cammalleri, I.; Gennaro, M.C.; La Bella, S. A survey of wild plant species for food use in Sicily (Italy)—Results of a 3-year study in four Regional Parks. *J. Ethnobiol. Ethnomed.* **2016**, *12*. [CrossRef]
3. Geraci, A.; Amato, F.; Di Noto, G.; Bazan, G.; Schicchi, R. The wild taxa utilized as vegetables in Sicily (Italy): A traditional component of the Mediterranean diet. *J. Ethnobiol. Ethnomed.* **2018**, *14*, 1–27. [CrossRef] [PubMed]
4. Lentini, F.; Venza, F. Wild food plants of popular use in Sicily. *J. Ethnobiol. Ethnomed.* **2007**, *3*, 1–12. [CrossRef] [PubMed]
5. Mattirolo, O.; Gallino, B.; Pallavicini, G. *Phytoalimurgia Pedemontana.*; Blu edizioni: Pevegnano, Italy, 2011; ISBN 9788879041218.
6. Elia, A.; Santamaria, P. Biodiversity in vegetable crops, a heritage to save: The case of Puglia region. *Ital. J. Agron.* **2013**, *8*, 4. [CrossRef]
7. Pieroni, A.; Nebel, S.; Santoro, R.F.; Heinrich, M. Food for two seasons: Culinary uses of non-cultivated local vegetables and mushrooms in a south Italian village. *Int. J. Food Sci. Nutr.* **2005**, *56*, 245–272. [CrossRef]
8. Morales, P.; Ferreira, I.C.F.R.; Carvalho, A.M.; Sánchez-Mata, M.C.; Cámara, M.; Fernández-Ruiz, V.; Pardo-de-Santayana, M.; Tardío, J. Mediterranean non-cultivated vegetables as dietary sources of compounds with antioxidant and biological activity. *LWT Food Sci. Technol.* **2014**, *55*, 389–396. [CrossRef]
9. Benincasa, P.; Tei, F.; Rosati, A. Plant density and genotype effects on wild asparagus (*Asparagus acutifolius* L.) spear yield and quality. *HortScience* **2007**, *42*, 1163–1166. [CrossRef]
10. Dorais, M.; Papadopoulos, A.P.; Luo, X.; Leonhart, S.; Gosselin, A.; Pedneault, K.; Angers, P.; Gaudreau, L. Soilless greenhouse production of medicinal plants in north Eastern Canada. *Acta Hortic.* **2001**, *554*, 297–303. [CrossRef]
11. Bonasia, A.; Conversa, G.; Lazzizera, C.; La Rotonda, P.; Elia, A. Weed control in lampascione - *Muscari comosum* (L.) Mill. *Crop Prot.* **2012**, *36*, 65–72. [CrossRef]
12. Branca, F.; Fisichella, A. Response of *Brassica fruticulosa* Cyr. to greenhouse cultivation. In Proceedings of the VI International Symposium on Protected Cultivation in Mild Winter Climate: Product and Process Innovation 614, Ragusa, Sicily, 8 March 2003; Volume 614, pp. 89–93.
13. Senejoux, F.; Demougeot, C.; Cuciureanu, M.; Miron, A.; Cuciureanu, R.; Berthelot, A.; Girard-Thernier, C. Vasorelaxant effects and mechanisms of action of *Heracleum sphondylium* L. (Apiaceae) in rat thoracic aorta. *J. Ethnopharmacol.* **2013**, *147*, 536–539. [CrossRef] [PubMed]
14. *Heracleum sphondylium* L.—USDA Plants Database. Available online: https://plants.sc.egov.usda.gov/home/plantProfile?symbol=HESP6 (accessed on 24 December 2022).
15. Borscht. Available online: https://en.wikipedia.org/wiki/Borscht (accessed on 25 January 2023).
16. Işcan, G.; Demirci, F.; Kürkçüoğlu, M.; Kivanç, M.; Başer, K.H.C. The bioactive essential oil of *Heracleum sphondylium* L. subsp. ternatum (Velen.) Brummitt. *Zeitschrift fur Naturforsch. Sect. C J. Biosci.* **2003**, *58*, 195–200. [CrossRef]
17. Uysal, A.; Ozer, O.Y.; Zengin, G.; Stefanucci, A.; Mollica, A.; Picot-Allain, C.M.N.; Mahomoodally, M.F. Multifunctional approaches to provide potential pharmacophores for the pharmacy shelf: *Heracleum sphondylium* L. subsp. *ternatum* (Velen.) Brummitt. *Comput. Biol. Chem.* **2019**, *78*, 64–73. [CrossRef]
18. Tranfield, D.; Denyer, D.; Smart, P. Towards a Methodology for Developing Evidence-Informed Management Knowledge by Means of Systematic Review. *Br. J. Manag.* **2003**, *14*, 207–222. [CrossRef]
19. *Heracleum sphondylium*. Available online: https://en.wikipedia.org/wiki/Heracleum_sphondylium (accessed on 26 January 2023).
20. *Heracleum sphondylium*: Sistematica, Etimologia, Habitat, Coltivazione. Available online: https://antropocene.it/2021/05/08/heracleum-sphondylium/ (accessed on 27 December 2022).
21. Devillers, P.; Devillers-Tcrschuren, J.; Ledant, J.-P. *CORINE Biotopes Manual. Habitats of the European Community. Data Specifications—Part 2.*; Office for Official Publications of the European Communities: Luxemburg, 1991; ISBN 92-826-3228-8.

22. Sheppard, A.W. *Heracleum sphondylium* L. *J. Ecol.* **1991**, *79*, 235–258. [CrossRef]
23. Dominik, T.; Pachlewski, R. Badanie mykotrofizmu zespolow roslinnych regla dolnego w Tatrac. *Acta Soc. Bot. Corum Pol.* **1956**, *25*, 3–26. [CrossRef]
24. Common Hogweed—Identification, Edibility, Distribution—Galloway Wild Foods. Available online: https://gallowaywildfoods.com/hogweed/ (accessed on 24 December 2022).
25. *Euleia heracleid.* Available online: https://www.eakringbirds.com/eakringbirds4/insectinfocuseuleiaheraclei.htm (accessed on 24 December 2022).
26. Anderson, H.A. *Berries: A Global History*; Reaktion Books: New York, NY, USA; Dover, DE, USA, 2018; ISBN 9781780239385.
27. Colombo, M.L.; Luciano, R. *Ombrellifere Della Provincia di Cuneo*; ArabaFenice, 2008; ISBN 9788895853000. Available online: https://www.amazon.co.uk/Ombrellifere-della-provincia-Cuneo-Colombo/dp/8895853008 (accessed on 29 January 2023).
28. Bussmann, R.W. *Ethnobotany of the Caucasus*; Bussmann, R.W., Ed.; Springer: Cham, Switzerland, 2017; ISBN 9783-319494111.
29. Renna, M.; Gonnella, M. The use of the sea fennel as a new spice-colorant in culinary preparations. *Int. J. Gastron. Food Sci.* **2012**, *1*, 111–115. [CrossRef]
30. Renna, M.; Gonnella, M.; Caretto, S.; Mita, G.; Serio, F. Sea fennel (*Crithmum maritimum* L.): From underutilized crop to new dried product for food use. *Genet. Resour. Crop Evol.* **2017**, *64*, 205–216. [CrossRef]
31. Bicchi, C.; D'Amato, A.; Frattini, C.; Cappelletti, E.M.; Caniato, R.; Filippini, R. Chemical diversity of the contents from the secretory structures of *Heracleum sphondylium* subsp. *sphondylium*. *Phytochemistry* **1990**, *29*, 1883–1887. [CrossRef]
32. Muckensturm, B.; Duplay, D.; Robert, P.C.; Simonis, M.T.; Kienlen, J.C. Substances antiappétantes pour insectes phytophages présentes dans *Angelica sylvestris* et *Heracleum sphondylium*. *Biochem. Syst. Ecol.* **1981**, *9*, 289–292. [CrossRef]
33. Brazdovicova, B.; Kostalova, D.; Strocinska, E.; Tomko, J. Isolation and identification of oroselol and other furocoumarin derivatives from *Heracleum sphondylium* L. roots. *Cesk. Farm.* **1982**, *31*, 346–347. [PubMed]
34. Maggi, F.; Quassinti, L.; Bramucci, M.; Lupidi, G.; Petrelli, D.; Vitali, L.A.; Papa, F.; Vittori, S. Composition and biological activities of hogweed [*Heracleum sphondylium* L. subsp. *ternatum* (Velen.) Brummitt] essential oil and its main components octyl acetate and octyl butyrate. *Nat. Prod. Res.* **2014**, *28*, 1354–1363. [CrossRef]
35. Melough, M.M.; Cho, E.; Chun, O.K. Furocoumarins: A review of biochemical activities, dietary sources and intake, and potential health risks. *Food Chem. Toxicol.* **2018**, *113*, 99–107. [CrossRef]
36. Paathak, M.A.; Farrington, D.; Fitzpatrick, T.B. The presently known distribution of furocoumarins (psoralens) in plants. *J. Investig. Dermatol.* **1962**, *39*, 225–239. [CrossRef] [PubMed]
37. Bahadori, M.B.; Dinparast, L.; Zengin, G. The Genus *Heracleum*: A Comprehensive Review on Its Phytochemistry, Pharmacology, and Ethnobotanical Values as a Useful Herb. *Compr. Rev. Food Sci. Food Saf.* **2016**, *15*, 1018–1039. [CrossRef] [PubMed]
38. Iscan, G.; Ozek, T.; Ozek, G.; Duran, A.; Baser, K.H.C. Essential oils of three species of *Heracleum*. Anticandidal activity. *Chem. Nat. Compd.* **2004**, *40*, 544–547. [CrossRef]
39. Fierascu, R.C.; Padure, I.M.; Avramescu, S.M.; Ungureanu, C.; Bunghez, R.I.; Ortan, A.; Dinu-Pirvu, C.; Fierascu, I.; Soare, L.C. Preliminary assessment of the antioxidant, antifungal and germination inhibitory potential of *Heracleum sphondylium* L. (Apiaceae). *Farmacia* **2016**, *64*, 403–408.
40. Benedec, D.; Hanganu, D.; Filip, L.; Oniga, I.; Tiperciuc, B.; Olah, N.K.; Gheldiu, A.M.; Raita, O.; Vlase, L. Chemical, antioxidant and antibacterial studies of Romanian *Heracleum sphondylium*. *Farmacia* **2017**, *65*, 252–256.
41. Baskin, C.C.; Baskin, J.M. *Seeds: Ecology, Biogeography, and Evolution of Dormancy and Germination*; Academic Press Inc.: San Diego, CA, USA, 1998; ISBN 978-0-12-080260-9.
42. Stokes, P. A Physiological Study of Embryo Development in *Heracleum sphondylium* L. *Ann. Bot.* **1952**, *16*, 441–447. [CrossRef]
43. Luna, T. Propagation Protocol for Production of Container (Plug) *Heracleum maximum* Bartr. Plants 172 mL Conetainers. Available online: https://npn.rngr.net/renderNPNProtocolDetails?selectedProtocolIds=apiaceae-heracleum-5 (accessed on 29 January 2023).
44. Nowruzian, A.; Masoumian, M.; Ebrahimi, M.A.; Bakhshi Khaniki, G.R. Effect of Breaking Dormancy Treatments on Germination of *Ferula assa-foetida* Seed. *Iran. J. Seed Res.* **2017**, *3*, 155–169. [CrossRef]
45. Renna, M. Reviewing the prospects of sea fennel (*Crithmum maritimum* L.) as emerging vegetable crop. *Plants* **2018**, *7*, 92. [CrossRef]
46. William, E. *Botanical Composition of the Park Grass Plots at Rothamsted 1856–1976*; Rothamsted Experimental Station: Hertfordshire, UK, 1978.
47. Fracchiolla, M.; Renna, M.; D'Imperio, M.; Lasorella, C.; Santamaria, P.; Cazzato, E. Living Mulch and Organic Fertilization to Improve Weed Management, Yield and Quality of Broccoli Raab in Organic Farming. *Plants* **2020**, *9*, 177. [CrossRef] [PubMed]
48. Fracchiolla, M.; Renna, M.; Durante, M.; Mita, G.; Serio, F.; Cazzato, E. Cover crops and manure combined with commercial fertilizers differently affect yield and quality of processing tomato (*Solanum lycopersicum* L.) organically grown in puglia. *Agriculture* **2021**, *11*, 757. [CrossRef]
49. Mudge, K.; Gabriel, S. *Farming the Woods: An Integrated Permaculture Approach to Growing Food and Medicinals in Temperate Forests*; Chelsea Green Publishing: White River Junction, VT, USA, 2014.
50. Bin, Y.W.; Hernandez, J.O.; Park, B.B. Effects of shade and planting methods on the growth of *Heracleum moellendorffii* and adenophora divaricata in different soil moisture and nutrient conditions. *Plants* **2021**, *10*, 2203. [CrossRef]

51. Henry, P.; Provan, J.; Goudet, J.; Guisan, A.; Jahodová, Š.; Besnard, G. A set of primers for plastid indels and nuclear microsatellites in the invasive plant *Heracleum mantegazzianum* (Apiaceae) and their transferability to *Heracleum sphondylium*. *Mol. Ecol. Resour.* **2008**, *8*, 161–163. [CrossRef] [PubMed]
52. Grzędzicka, E. Invasion of the Giant Hogweed and the Sosnowsky's Hogweed as a Multidisciplinary Problem with Unknown Future—A Review. *Earth* **2022**, *3*, 287–312. [CrossRef]
53. Niinikoski, P.; Korpelainen, H. Population genetics of the invasive giant hogweed (*Heracleum* sp.) in a northern European region. *Plant Ecol.* **2015**, *216*, 1155–1162. [CrossRef]
54. Anaclerio, M.; Renna, M.; di Venere, D.; Sergio, L.; Santamaria, P. Smooth golden fleece and prickly golden fleece as potential new vegetables for the ready-to-eat production chain. *Agriculture* **2021**, *11*, 74. [CrossRef]
55. Montesano, F.F.; Gattullo, C.E.; Parente, A.; Terzano, R.; Renna, M. Cultivation of Potted Sea Fennel, an Emerging Mediterranean Halophyte, Using a Renewable Seaweed-Based Material as a Peat Substitute. *Agriculture* **2018**, *8*, 96. [CrossRef]
56. Gonnella, M.; Renna, M. The evolution of soilless systems towards ecological sustainability in the perspective of a circular economy. Is it really the opposite of organic agriculture? *Agronomy* **2021**, *11*, 950. [CrossRef]
57. D'Imperio, M.; Parente, A.; Montesano, F.F.; Renna, M.; Logrieco, A.F.; Serio, F. Boron Biofortification of *Portulaca oleracea* L. through Soilless Cultivation for a New Tailored Crop. *Agronomy* **2020**, *10*, 999. [CrossRef]
58. D'Imperio, M.; Durante, M.; Gonnella, M.; Renna, M.; Montesano, F.F.; Parente, A.; Mita, G.; Serio, F. Enhancing the nutritional value of *Portulaca oleracea* L. by using soilless agronomic biofortification with zinc. *Food Res. Int.* **2022**, *155*, 111057. [CrossRef] [PubMed]
59. D'Imperio, M.; Renna, M.; Cardinali, A.; Buttaro, D.; Santamaria, P.; Serio, F. Silicon biofortification of leafy vegetables and its bioaccessibility in the edible parts. *J. Sci. Food Agric.* **2016**, *96*, 751–756. [CrossRef]
60. Puccinelli, M.; Pezzarossa, B.; Pintimalli, L.; Malorgio, F. Selenium biofortification of three wild species, *Rumex acetosa* L., *Plantago coronopus* L., and *Portulaca oleracea* L., grown as microgreens. *Agronomy* **2021**, *11*, 1155. [CrossRef]
61. Nascimento, C.S.; Nascimento, C.S.; Lopes, G.; Carrasco, G.; Lupino, P.; Bernardes, A. Biofortified Rocket (*Eruca sativa*) with Selenium by Using the Nutrient Film Technique. *Horticulturae* **2022**, *8*, 1088. [CrossRef]
62. Renna, M.; D'Imperio, M.; Maggi, S.; Serio, F. Soilless biofortification, bioaccessibility, and bioavailability: Signposts on the path to personalized nutrition. *Front. Nutr.* **2022**, *9*, 1–16. [CrossRef]
63. Buttaro, D.; Renna, M.; Gerardi, C.; Blando, F.; Santamaria, P.; Serio, F. Soilless production of wild rocket as affected by greenhouse coverage with photovoltaic modules. *Acta Sci. Pol. Cultus.* **2016**, *15*, 129–142.
64. Renna, M.; Montesano, F.; Gonnella, M.; Signore, A.; Santamaria, P. BiodiverSO: A Case Study of Integrated Project to Preserve the Biodiversity of Vegetable Crops in Puglia (Southern Italy). *Agriculture* **2018**, *8*, 128. [CrossRef]

MDPI
St. Alban-Anlage 66
4052 Basel
Switzerland
www.mdpi.com

Horticulturae Editorial Office
E-mail: horticulturae@mdpi.com
www.mdpi.com/journal/horticulturae

www.ingramcontent.com/pod-product-compliance
Lightning Source LLC
Chambersburg PA
CBHW062034220926
51850CB00010B/2549
* 9 7 8 3 0 3 6 5 9 6 7 0 9 *